Studies in Computational Intelligence

Volume 670

Series editor

About this Series

The series "Studies in Computational Intelligence" (SCI) publishes new developments and advances in the various areas of computational intelligence—quickly and with a high quality. The intent is to cover the theory, applications, and design methods of computational intelligence, as embedded in the fields of engineering, computer science, physics and life sciences, as well as the methodologies behind them. The series contains monographs, lecture notes and edited volumes in computational intelligence spanning the areas of neural networks, connectionist systems, genetic algorithms, evolutionary computation, artificial intelligence, cellular automata, self-organizing systems, soft computing, fuzzy systems, and hybrid intelligent systems. Of particular value to both the contributors and the readership are the short publication timeframe and the worldwide distribution, which enable both wide and rapid dissemination of research output.

More information about this series at http://www.springer.com/series/7092

Quan Bai · Fenghui Ren
Katsuhide Fujita · Minjie Zhang
Takayuki Ito
Editors

Multi-agent and Complex Systems

Springer

Editors
Quan Bai
Auckland University of Technology
Auckland
New Zealand

Minjie Zhang
University of Wollongong
Wollongong, NSW
Australia

Fenghui Ren
University of Wollongong
Wollongong, NSW
Australia

Takayuki Ito
Nagoya Institute of Technology
Nagoya
Japan

Katsuhide Fujita
Tokyo University of Agriculture and Technology
Koganei, Tokyo
Japan

ISSN 1860-949X ISSN 1860-9503 (electronic)
Studies in Computational Intelligence
ISBN 978-981-10-2563-1 ISBN 978-981-10-2564-8 (eBook)
DOI 10.1007/978-981-10-2564-8

Library of Congress Control Number: 2016951707

Printed on acid-free paper

This Springer imprint is published by Springer Nature
The registered company is Springer Nature Singapore Pte Ltd.
The registered company address is: 152 Beach Road, #22-06/08 Gateway East, Singapore 189721, Singapore

Preface

Computer-based modelling and simulation have become useful tools to facilitate understanding of systems in diverse domains such as physics, astrophysics, chemistry, biology, economics, engineering, and social science. A complex system is characterized by a large number of interacting components (e.g. agents and processes) whose aggregate activities are nonlinear and self-organized. Complex systems are hard to simulate or model using traditional computational approaches because of the complex relationships of components and distributed features of resources and dynamic work environments. Meanwhile, smart systems such as multi-agent systems have demonstrated advantages and great potential in modelling and simulating complex systems.

The International Workshop on Smart Simulation and Modelling for Complex Systems (SSMCS'15) was held in Buenos Aires, Argentina, in July 2015, and the International Joint Agents Workshop and Symposium (IJAWS'15) was convened in Ishikawa, Japan, in October 2015. The aims of SSMCS'15 and IJAWS'15 were to bring together researchers in artificial intelligence (AI), agent and multi-agent systems, and system modelling/simulation to discuss research challenges and cutting-edge techniques in smart simulation and modelling. This book contains the extended versions of selected papers from the SSMCS'15 and IJAWS'15 workshops. For those workshops we solicited papers on all aspects of smart simulation and modelling of complex systems through the use of agent and other AI technologies. They are, for instance, being studied in network modelling, microsimulation modelling, social influence modelling, disaster modelling, environment modelling, power market modelling and idea-discovery-process modelling. The goal of the workshops was to gather researchers from these communities to learn from one another, form long-term collaborations and cross-fertilize the various disciplines to accelerate progress toward more complex and realistic applications.

Finally, we would like to extend our sincere thanks to all authors. This book would not have been possible without the valuable support and contributions of those who cooperated with us.

Auckland, New Zealand — Quan Bai
Wollongong, Australia — Fenghui Ren
Koganei, Japan — Katsuhide Fujita
Wollongong, Australia — Minjie Zhang
Nagoya, Japan — Takayuki Ito
May 2016

Contents

Part II Intelligent Agents and Their Applications

Part I
Smart Simulation and Modelling

Chapter 1
Simulating and Modeling Dual Market Segmentation Using PSA Framework

Jiamou Liu, Ziheng Wei and Quan Bai

Abstract Market segmentation refers to the analytical process of dividing a broad market into segments taking into account multiple factors such as consumer needs, interests and tastes; it has been considered one of the most important marketing strategies as it helps a business to identify hidden market trends, define target segments, and design marketing plans. Market segmentation may also be viewed as a computational challenge: Given the massive amount of data describing interactions between consumers and commodities, the task is to partition the set of consumers and commodities into subsets that corresponds to market segments—two consumers are in the same segments when they exhibit a similar purchasing pattern, while two products are in the same segments when they are purchased by a similar group of consumers. In this work, we focus on the definition and simulation of market segments. We employ the Propose-Select-Adjust (PSA) framework, introduced in an earlier work [10], to simulate the forming of market segments. Our approach is distributed and can be applied to large and dynamic market data set. The experimental results suggest that the proposed approach is a promising technique for supporting intelligent market segmentation.

Keywords Market segmentation · PSA framework · Bipartite network

J. Liu (✉) · Z. Wei
Department of Computer Science,
The University of Auckland, Auckland, New Zealand
e-mail: jiamou.liu@auckland.ac.nz

Z. Wei
e-mail: zwei891@aucklanduni.ac.nz

Q. Bai
School of Engineering, Computer and Mathematical Sciences,
Auckland University of Technology, Auckland, New Zealand
e-mail: quan.bai@aut.ac.nz

Q. Bai et al. (eds.), *Multi-agent and Complex Systems*, Studies in Computational Intelligence 670, DOI 10.1007/978-981-10-2564-8_1

1.1 Introduction

"I Shop Therefore I Know That I Am", stated British sociologist Colin Campbell in a 2004 research article; this sentence has become a motto of modern consumerism: people increasingly justify their existence and identify themselves by the products they shop [1]. Indeed, the theory of "social groups" asserts that consumption is intimately tied to the question of "with whom do I belong" and the formation of collective identities—the so-called *social groups*; they capture notions such as trends, style and sub-cultures [13]. Analysing social groups helps us to decode the hidden patterns of consumer behaviours, differentiate sophisticated broad markets and identify trends and dynamics within them.

From a marketer's perspective, understanding the market amounts to understand the desire, preference and priorities of consumers; this task is called *market segmentation*. Market segmentation is one of the most important marketing strategies. Its aim is to divide a broad market into smaller groups, taking into account different factors such as customers' interests, and tastes, revealing important information regarding lifestyles, geographic differences, and other demographic and economical phenomena. With this information, a business defines its target customers, enabling more accurate orientation of marketing plans, allowing more efficient use of resources and hence increasing profits.

Market segmentation is a complex and data intensive task. In order to pinpoint the exact correlation between the consumption behaviours and consumers' life styles and trends, analysts need to process and catagorise a huge amount of market data such as shopping records. In the past 5 years, there has been a significant interest in computational approaches that support automated market segmentation and the problem has become a central problem in data mining [3, 11, 12, 14]. One common goal of these works is to build an intelligent tool which, through analysing large amounts of market data, computes an appropriate segmentation of the market.

Research on market segmentation exhibits a diverse landscape. For example, Miguis et al. recently use lifestyle information derived from market segmentation to analyse customer purchasing power [11]. In another work, the same authors also investigate the design of product promotion strategies based on market segmentation results [12]. Market segmentation has also been applied in customer loyalty management [14].

We point out here that market segmentation can be viewed from two perspectives: the *consumers*' perspective and the *commodities*' perspective. On one hand, market segmentation aims to categorise consumers into different social groups. For example, two consumers may be characterised as having a similar purchasing pattern and fall into the same social group. On the other hand, the process also aim to distinguish clusters of commodities. For example, two products may be seen as correlated as they are likely to be purchased by the same group of consumers. These two perspectively naturally influence each other, as the forming of a consumer group may help to identify a new commodity segment, and the introduction of a range of products may also enforce the emergence of a consumer group. The relation between consumers and commodities are interactive, highly dynamic and often subtle. It is therefore an important question to analyse interactions between consumers and commodities.

Based on the above observations, we propose the *dual market segmentation problem*. The underlying framework treats a market as a bipartite graph consisting of consumer nodes and commodity nodes, and interactions between them. The goal of the problem is to design a computational approach for segmenting—at the same time—both the consumers and commodities, taking into account these interactions.

In our previous work [9], we proposed a decentralised computational framework, Propose-Select-Adjust (PSA), for solving network problems. In this framework, each node of the graph acts as a individual computational unit which performs three procedures in cyclic order: Propose, Select and Adjust. The nodes are acting in an asynchronous manner while each node only has access to its local information. Through a series of case studies and experiments, we demonstrated the approach is a promising solution for simulating community formation, and hence detecting communities in a large dynamic network [10]. In this paper, we extend the PSA framework on the bipartite market network connecting consumers and commodities. We suggest that PSA can be applied to market segmentation. This approach has several advantages:

- Firstly, PSA is a technology for smart simulation of group behaviors in a network. Thus it simulates the formation of segments within the consumers and commodities. This simulation naturally captures the mutual influence between consumers and commodities, which is otherwise hard to define.
- Secondly, PSA as a distributed framework, can scale to handle large amount market data, while computation is kept local.
- Thirdly, as a PSA cell runs continuously, the PSA cells monitors the changing market, and detects emergence of market segments.

We support our claims above with experimental results on both synthesised and real data.

The rest of the paper is organised as follows. Section 1.2 formulates the dual market segmentation problem using a bipartite graph model. Section 1.3 describes the PSA framework and extend it to the bipartite dual network of consumers and commodities. Our new framework provides two perspectives: The *local perspective* segments consumers and commodities based on their links. The *global perspective*, on the other hand, captures the interactions between the segments identified in the local perspective. Our experimental results are presented and discussed in Sect. 1.4. Some related works are reviewed in Sect. 1.5. Finally the paper is concluded in Sect. 1.6.

1.2 Problem Formulation

A *bipartite graph* is a pair $(V_0 \cup V_1, E)$ where V_0 and V_1 are two sets of nodes with $V_0 \cap V_1 = \varnothing$, and the edge relation E is a subset of the Cartesian product $V_0 \times V_1$. We use bipartite graph to abstractly model transactions in a market. In particular, we may view V_0 as a set of *consumers*, and V_1 as a set of *commodities* in a market; an edge $(v, p) \in V_0 \times V_1$ denote that consumer v purchases the commodity p. Hence this bipartite graph captures a collection of purchasing records of all consumers in V_0. We point out that the terms "consumers" and "commodities" are used in a broad sense: these are two abstract concepts which captures any interacting objects in a market-like context.

This bipartite graph model involves a *dual network*: On one hand, from the perspective of a consumers, its outgoing edges indicate its *purchased commodities*—intuitively, two consumers have similar purchasing behaviours if they purchase similar commodities. On the other hand, from the perspective of a commodity, its incoming edges indicate its *consumer base*—intuitively, two commodities have similar consumer base, if they are purchased by similar groups of customers.

The main problem of the paper is to compute a segmentation of this dual network. For consumers, a segmentation amounts to dividing the market into groups of consumers with similar purchasing patterns which are affected by consumers' tastes, life styles, desires, etc. For commodities, a segmentation amounts to aggregation of correlated products on the market. Formally, we model market segmentation as follows:

Definition 1 (*Dual segmentation*) Let $G = (V_0 \cup V_1, E)$ be a bipartite graph. A *dual segmentation* is a pair $(\sim_0, \sim_1)$ where $\sim_i$ is an equivalence relation on V_i for $i \in \{0, 1\}$. An $\sim_i$-equivalence class is called a V_i *-segment.*

For any consumer $v \in V_0$, the V_0-segment of v contains all consumers $u \in V_0$ such that $v \sim_0 u$; for any commodity $x \in V_1$, the V_1-segment of x contains all products $y \in V_1$ such that $x \sim_1 y$. We observe that, in the context of a market, the two components of a dual segmentation are interrelated. For example, consumers with a similar lifestyle tend to purchase a similar group of products, which in turn causes correlation among these products. Therefore, our goal is to provide a uniform approach for computing a dual segmentation, in which the segments of consumers influence segments of commodities, and vice versa.

This approach builds the two networks involved in the market: the consumer network, and the commodity network. A *similarity function* on a set S is a function $\omega : S \times S \to \mathbb{R}$. The key ingredients of our approach are two similarity functions, ω_0, ω_1, that are defined on the two sets V_0 and V_1, respectively. Our similarity functions ω_0, ω_1 are based on *Jaccard similarity*, which is one of the most commonly used similarity measure for sets. Given a bipartite graph $G = (V_0 \cup V_1, E)$, for $u \in V_0$, $v \in V_1$, define

$$E(u) = \{v' \in V_1 \,|\, (u, v') \in E\} \text{ and } E^{-1}(v) = \{u' \in V_0 \,|\, (u', v) \in E\}.$$

The similarity $\omega_0(v_1, v_2)$ between $v_1, v_2 \in V_0$ is the Jaccard similarity

$$\omega_0(v_1, v_2) = \frac{|E(v_1) \cap E(v_2)|}{|E(v_1) \cup E(v_2)|} \tag{1.1}$$

The similarity $\omega_1(v_1, v_2)$ between $v_1, v_2 \in V_1$ is the Jaccard similarity

$$\omega_1(v_1, v_2) = \frac{|E^{-1}(v_1) \cap E^{-1}(v_2)|}{|E^{-1}(v_1) \cup E^{-1}(v_2)|} \tag{1.2}$$

Using these two similarity functions, two networks can be constructed:

Definition 2 *(V_i-networks)* Let $G = (V_0 \cup V_1, E)$ be a bipartite graph, and ω_0, ω_1 be two similarity functions on V_0 and V_1 respectively. For any $i \in \{0, 1\}$, the *V_i-network* is an undirected graph (V_i, E_i) where

$$E_i = \left\{(u_1, u_2) \in V_i^2 \,|\omega_i(u_1, u_2) \geq \alpha_i \right\}$$

where the constants α_i is called the *local similarity threshold.*

We call V_0-network and V_0-segments as *consumer network* and *consumer segments*, and call V_1-network and V_1-segments as *commodity network* and *commodity segments*, respectively.

In the rest of the paper, we describe a simulation of market segmentation in two steps: *local segmentation* and *global segmentation*. For local segmentation, we construct for each $i \in \{0, 1\}$, a set of disjoint V_i-segments $\mathscr{C}_i = \{C_{1,i}, ..., C_{\ell,i}\}$ where each $C_{j,i} \subseteq V_i$ and $\bigcup C_j \in \mathscr{C}_i = V_i$. Intuitively each $C_{j,0}$ represents a group of consumers who share some common interests. Thus each V_i-segment ideally should be a maximal clique (i.e., a complete subgraph) in the V_i-network. Since finding maximal cliques is a well-known NP-hard problem, we use the following notions to approximate a clique-like subgraph.

Definition 3 *(Local core)* In a graph $G = (V, E)$, $N(v)$, the *closed neighbourhood* of v, is a set $\{u \,|(u, v) \in E\} \cup \{v\}$ of all nodes that are adjacent to v. A *k-core* in a graph is an induced subgraph where all nodes have degree at least k. The *core number* of a node v is the largest $\kappa(v)$ such that $N(v)$ contains a $\kappa(v)$-core. For $v \in V$, the *local core* of v is the set

$$K(v) = \{u \in N(v) \,||N(u) \cap N(v)| \geq \kappa(v) - 1\}$$

Intuitively, the local core of a node v is a subgraph that contains those nodes adjacent to v and are adjacent to at least $\kappa(v)$ many other nodes in the local core. The value of $\kappa(v)$ measure how "tightly-knitted" this subgraph is; when $\kappa(v)$ equals to the size of the local core minus 1, the local core is a clique.

The output of local segmentation consists of a set $\mathscr{C}_0$ of consumer segments and a set $\mathscr{C}_1$ of commodity segments. From the perspective of consumers, several consumer segments can share common interests towards the same commodity segments. From the perspective of commodities, several commodity segments are deployed to the same consumer segments. This calls for possible combinations of these already computed segments. Therefore, in global segmentation, we take the edge set $E(C_1, C_2) = \{(u, v) \in C_0 \times C_1 \,|(u, v) \in E\}$, and construct a bipartite graph $G^H = (\mathscr{C}_0 \cup \mathscr{C}_1, E^H)$ where

$$E^H = \left\{(C_1, C_2) \in \mathscr{C}_i \times \mathscr{C}_{1-i} \left| \frac{|E(C_1, C_2)|}{|C_1|} \geq \beta_i, i \in \{0, 1\}\right.\right\}$$

where β_i is the *segmentation selection rate*, which determines the importance of the V_2-segment C_2 to the V_1-segment C_1.

Similarly to (1.1) and (1.2), we define *global similarity functions* ω_0^H and ω_1^H on $\mathscr{C}_0$ and $\mathscr{C}_1$, respectively. The graph G^H has corresponding dual networks: $\mathscr{C}_0$- and $\mathscr{C}_1$-network where

$$E_i^H = \{(C_1, C_2) \in \mathscr{C}_i^2 \,|\omega_i^H(C_1, C_2) \geq \gamma_i\}$$

where γ_i is the *global similarity threshold*. The similarity $\omega_i^H(C_1, C_2)$ between C_1, $C_2 \in \mathscr{C}_i$ where $i \in \{0, 1\}$ is the Jaccard similarity.

1.3 The PSA Framework for Segmentation Simulation

1.3.1 General Description

The Propose-Select-Adjust (PSA) framework is a distributed computational framework for simulating a network of interconnected agents, called *cells*, which perform some homogeneous operations in an asynchronous manner. The framework was suggested by the authors in [9] to simulate community formation and detection in a distributed and dynamic setting.

In this paper, we extend the Propose-Select-Adjust (PSA) framework to simulate market segmentation based on a bipartite graph model. Intuitively, each node in the bipartite graph is a cell which carries out certain tasks independently. The PSA framework describes how to implement a single cell. The framework is inspired by the decision making process among a group of people: Imagine a group of individuals trying to decide on a partitioning of the group, where every member would belong to one and only one subgroup. The constraint is that each individual only sees local information about her own connections; no knowledge is shared among all members. Thus individuals can only make self-centred judgements and decide on the people that she would like to be with. Under this constraint, the following procedures can ensure the group arriving at a collective decision:

- Propose: Each person independently writes down a list of people whom she would like to join to form her own subgroup. She then sends an invitation to everyone on the list to form a group. Here we implicitly assume that a person makes an invitation to herself.
- Select: During the Propose phase, a person would receive a number of invitations from its neighbours. After all invitations are received, the person evaluates the quality of each proposal, then selects and accepts the best proposal.
- Adjust: Once a person accepts an invitation, she then updates her own decision according to the accepted proposal. After every individual finishes this step, the whole group would have been divided into a number of subgroups, and thus a clustering is formed.

More precisely, the crucial ingredients in the definition of a PSA cell consist of a set of proposals and a preference relation, which are defined as follows:

Definition 4 (*Proposal and Preference*) A *proposal* of $v \in V$ is a set $\mathscr{P}_v \subseteq N(v)$. Thus $2^{N(v)}$ denotes the set of all possible proposals of v. A *preference relation* $\preceq$ is a linear ordering defined on all finite graphs such that for any two graph G_1, G_2, $G_1 \prec G_2$ means G_1 is preferred over G_2.

Any implementation of the PSA system involves giving a precise definition of proposals for each cell, and a preference relation. In our bipartite graph market model, each consumer and each commodity will act as a PSA cell, which carries out computation based on its local information: For a consumer this information contains all commodities that consumer chooses; for a commodity, this information contains all consumers that choose this commodity.

Each cell performs three phases of computation repeatedly, which are roughly described below: In the Propose phase, every cell v prepares and announces a proposal $\P_v$ which is received by all nodes contained in the proposal. In particular, as the proposal $\P_v$ will automatically contain the node v itself, we also regard v as having received the proposal. Then in the Select phase, every cell v takes the collection of proposals it receives, and chooses a most preferred proposal according to the relation $\preceq$. Suppose v selects the proposal $\P_u$ proposed by u. In the Adjust phase, v observes the action of u and makes the two possible actions:

1. If u chooses its own proposal $\P_u$ (in this case, its own proposal is seen as the most preferred option), then v joins the group of u.
2. If u chooses the proposal $\P_w$ of some node $w \neq u$, then v does not join the group of u, and disregard the proposal from u.

After finishing the Adjust phase, every cell would go back to the Propose phase and restart the three phases again. This enables computation in a dynamic setting: Suppose the input data (in the form of edges) is dynamic, every cell would make spontaneous changes to recompute new proposals every time it reaches the Propose phase; then its neighbours will modify their selections and adjust their solutions correspondingly; such changes may cascade through the graph. Moreover, assuming the market data is stable (so no change occurs anymore), every PSA cell in this simulation will eventually reach a stable solution.

To implement a PSA system to solve the dual market segmentation problem on a bipartite graph, we must handle both the local and global segmentation steps. Therefore, we need one PSA system to simulate local segmentation of consumers and commodities, and another PSA system to simulate global segmentation. Hence, we use two types of PSA cells: local PSA cells and global PSA cells.

1.3.2 *Local* PSA *Cells*

In a system $G = (V_0 \cup V_1, E)$, we compute the corresponding V_0-network (V_0, E_0) and V_1-network (V_1, E_1) as described above. In an undirected graph, a local core represents a set of nodes that are similar to each other. Therefore each cell $v \in V$ makes a proposal $\mathscr{P}_v = K(v)$. In the graph $G = (V, E)$, for any set $C \subseteq V$, we use $N(C)$ to denote the union $\bigcup_{u \in C} N(u)$. We utilise the following density measures:

- The *intra cluster density* of C is the percentage of the number of edges in C over all possible internal edges; in other words, we define

$$d_{intra}(C) = \frac{2 \times |E \upharpoonright C|}{|C| \times (|C| - 1)} \tag{1.3}$$

- The *inter cluster density* is the percentage of the number of edges connecting C with an outside node over all possible links from C to outside nodes; in other words, we define

$$d_{inter}(C) = \frac{|(E \upharpoonright N(C)) \setminus (E \upharpoonright C)|}{|C| \times |N(C) \setminus C|} \tag{1.4}$$

Combining these two factors, we obtain the *utility* function d defined on a set C of nodes:

$$d(C) = d_{intra}(C) \times (\kappa(v) + 1) - d_{inter}(C) \times |C|$$

Note that intuitively, a set C has higher utility $d(C)$ if it is dense, has relatively large cardinality, and sparsely connected to nodes outside of it. To define the *preference* relation between proposals, we set $C_1 \preceq C_2$ whenever $d(C_1) \geq d(C_2)$.

As described above, every cell v makes a proposal $\P_v$ to its neighbours in the Propose phase. In this propose every cell would eventually receive a certain number of proposals. Each cell then anlayses all its received proposals and it chooses a most preferred proposal according to the preference relation μ defined above.

1.3.3 *Global* PSA *Cells*

After constructing the local segmentation, we obtain a new bipartite graph

$$G^H = (\mathscr{C}_0 \cup \mathscr{C}_1, E^H)$$

whose nodes are segments computed by the local PSA cells. We use another set of PSA cells to compute a global segmentation, where each cell stands for a local segment in $\mathscr{C}_0 \cup \mathscr{C}_1$. Similarly to the local perspective, the global perspective also consists of a dual networks: $(\mathscr{C}_0, E_0^H)$ and $(\mathscr{C}_1, E_1^H)$. For each $C \in \mathscr{C}_i$ where $i \in \{0, 1\}$, the proposal of C is $\mathscr{P}_C = \{C', C\}$ such that for all C_i $(C_i, C) \in E_i^H$ we have $\omega_i^H(\mathscr{C}', \mathscr{C}) \geq \omega_i^H(\mathscr{C}_i, \mathscr{C})$. Essentially, whenever two segments have maximum global Jaccard similarity, this process would form a larger segment by combining these two segments.

1.4 Experiment

1.4.1 Metrics for Performance Evaluation

We describe the basic setup for evaluating the proposed approach. Our data set represents a bipartite graph $(V_0 \cup V_1, E)$, to which we apply our PSA system. For any consumer $v \in V_0$, we use $\mathscr{C}_v$ to denote the set of commodities that v is attracted to. In other words,

$$\mathscr{C}_v = E(v) = \{u \in V_1 \,|\{v, u\} \in E\}. \tag{1.5}$$

For our experiments, we fix a percentage $p \in [0, 1]$ and "hide" p percent of the edges in the data set. This results in a subgraph $(V_0 \cup V_1, E')$ where $E' \subseteq E$, called the *test data set*. The *known set* for the consumer v is $\mathscr{C}_v^{\mathsf{s}} = E'(v) = \{u \in V_1 \,\big|\{v, u\} \in E'\}$. Similarly, the known set for any commodity $u \in V_1$ is $\mathscr{C}_u^{\mathsf{s}} = E'^{-1}(u) = \{v \in V_0 \,\big|\{v, u\} \in E'\}$. For any $v \in V_0 \cup V_1$, the PSA system allocates the known set $\mathscr{C}_v^{\mathsf{s}}$ to the cell corresponding to u. The PSA system will produce, for any consumer $v \in V_0$ a segment $S(v) \subseteq V_0$ which contains v. We use $\mathscr{C}_v^{\mathsf{r}}$ to denote the set of commodities that are linked to all members of the segment $S(v)$. In other words,

$$\mathscr{C}_v^{\mathsf{r}} = \{u \,\big|\exists v' \in S(v) : \{u, v'\} \in E\}. \tag{1.6}$$

To validate our approach, we make the following assertion: In the ideal case, the market segments found by the approach should truthfully reflect consumers' interests in different products. This means, a desirable outcome of the market segmentation process, is that each consumer would have a strong interest in the products linked by all others in his own segment. This means that the desirable outcome is that $\mathscr{C}_v^{\mathsf{r}} = \mathscr{C}_v$. We will use this as a criterion for the validity of our PSA-based segmentation mechanism: We compare the set $\mathscr{C}_v^{\mathsf{r}}$ with the set $\mathscr{C}_v$ using three metrics: *precision*, *recall* and *successful deployment rate*. Precision and recall are two metric commonly used in data mining for demonstrating the performance of an automatic recommendation system [2].

- The precision of user v is defined as:

$$\mathsf{precision}(v) = \frac{|\mathscr{C}_v^{\mathsf{r}} \cap \mathscr{C}_v|}{|\mathscr{C}_v^{\mathsf{r}}|} \tag{1.7}$$

  Intuitively, the value of $\mathsf{precision}(v)$ is the percentage of "correctly identified" commodities within the all "identified" commodities. It captures the amount of commodities chosen by a consumer within the computed segment of this consumer. Naturally, the higher its value, the higher the degree of relevance between v and the commodity segment $\mathscr{C}_v^{\mathsf{r}}$. In particular, the precision of a commodity segmentation should be higher than the its *random selection rate* $\frac{|\mathscr{C}_v|}{|V_1|}$ (V_1 is a set of all commodities), which is the value if $\mathscr{C}_v^{\mathsf{r}}$ is chosen randomly.
- The recall of v is defined as:

$$\mathsf{recall}(v) = \frac{|\mathscr{C}_v^{\mathsf{r}} \cap \mathscr{C}_v|}{|\mathscr{C}_v|} \tag{1.8}$$

  The value of $\mathsf{recall}(v)$ is the percentage of "correctly identified" commodities within the total commodities "chosen by the consumer". Only considering precision may have some side effects. For example, in order to achieve high precision, a commodity may become conservative to include more commodities. Improving recall with segmentation of a commodity segmentation enforces it to include more attractive commodities. Similarly to precision, the *expected recall* is $\frac{|\mathscr{C}_v^t|}{|\mathscr{C}_v|}$.
- The successful deployment rate of v is:

$$\mathsf{sdr}(v) = \frac{|\mathscr{C}_v^{r} \cap (\mathscr{C}_v \setminus \mathscr{C}_v^{t})|}{|\mathscr{C}_v \setminus \mathscr{C}_v^{t}|} \tag{1.9}$$

  In other words, $\mathsf{sdr}(v)$ denotes the amount of products "unknown" to v that are linked to others in the same segments as v; This value represents how much "hidden" information can be "re-discovered" by the segmentation process. Hence a satisfying segmentation process should result in high values of $\mathsf{sdr}(v)$.

1.4.2 Data Set

We take the Jester data set [5], which contains Internet users' ratings of a collection of jokes. In this setting, an online user is a consumer and a joke is the corresponding commodity. Segmentation of this data set should reveal users' different tastes on humour. We select one of the Jester data sets which contains 24,983 users and 101 jokes. Each user rates at least 36 jokes by giving an integer value ranging from −10 to 10. We further processed the data set by assuming that a user likes a joke if its rating is non-negative. We aim to use PSA system to identify segmentations of people with similar humour taste.

Our approach should be able to reveal whether a user would like a given joke. To test the validity of the approach, we pick a test data set from the original data set by keeping, for each user, only 80 % of his rating of the jokes. We then apply our approach to process the test data set, which produces a number of segments. We test how much the results recovers the original rating (on all jokes).

Jester data set is an *imbalanced data set* because the number of consumers is much larger than the number of commodities. To understand the performance of our approach on *balanced data sets*, i.e., those where the number of consumers and number of commodities are both large, we design an experiment which involves a synthetic data set. In generating this synthetic data set, we use the following parameters:

- the number of consumers and commodities (σ_1, σ_2)
- the probability of selecting a segmentation in the dual network (σ_3)
- the probability of generating an edge to a node in a selected segmentation of the dual network (σ_4)
- the probability of generating an edge to a node in a segmentation which is not selected of the dual network (σ_5)

If two segmentations on the different sides of a dual network select each other, then the probability of generating edges between nodes in these two segmentations is high. Otherwise the probability should be low. Note that one segmentation has to select at least one segmentation and it also has a probability to select more than one segmentation. Our synthetic data set indeed makes sure that certain relationships are established between segmentations. In the experiment, we give all the edges to our system. Instead, our system needs to find all the segmentations and selected segmentations in the dual network. Our settings for a synthetic data set as follow:

- $\sigma_1 = \sigma_2 = 500$, $\sigma_3 = 0.5$, $\sigma_4 = 0.8$ $\sigma_5 = 0.2$
- $\alpha_0 = \alpha_1 = 0.7$, $\beta_0 = \beta_1 = 0.5$, $\beta_0 = \beta_1 = 0.6$.

1.4.3 Experiment Environment

The PSA framework is designed to be implemented in a distributed environment where cells functions autonomously. Due to physical limitation, we perform computation in a single workstation, but implement a number of parallel processes, each simulating a group of cells. In our simulation, each cell only has to process a small amount of local information surrounding it. So, this parallel framework greatly improves computation time. We implement PSA system using the Java programming language as a multi-threaded application. To improve efficiency, we evenly distribute all the cells in a data set into four parallel threads. Each thread is responsible to execute PSA cells sequentially. This process results in faster computation comparing to sequentially executing of all PSA cells in series. The detailed specification of our experiment environment is: Windows 7 operating system, Intel Core i5-4570 quad-core CPU 3.2 GHz and 16 GB RAM.

1.4.4 Experimental Results and Discussion

We conduct four experiments on the Jester data set. We use PSA system to identify segmentations of users with similar tastes for jokes. In Experiment 1, we take 10 disjoint test data sets consisting of 1000 consumers (each set contains a disjoint set of consumers) and 101 commodities each. For any consumer, we retain 80 % commodities it links to in the data set. We set $\alpha_0 = 0.35$, $\beta_0 = 0.3$, $\gamma_0 = 0.5$ and $\alpha_1 = 0.45$, $\beta_1 = 0.7$, $\gamma_1 = 0.2$. The results are shown in Fig. 1.1.

The Jester data set is imbalanced as the number of users greatly exceeds that of times. So we have to configure different parameters for each dual network. The random selection rate result shows that, in general, our approach has more than 40 % chance to correctly compute the segment of a consumer than making random guesses. The result shows that our approach improves the precision to 80 %. In addition, it finds almost 60 % of hidden commodities.

In *Experiment 2*, our goal is to analyse how the data set quality influences the results. We take consumers in *Dataset 1* from *Experiment 1* and generate 10 test data sets by varying the amount of links that are hidden from the test data set. The results show that with a higher percentage of deleted links, it is more difficult to compute correct segments. The results of *Experiment 2* are shown in Fig. 1.2.

Precision describes how less mistakes that a system makes. When we reduce the percentage of given commodities, our system starts to lose capability of finding correlated segmentations within current parameter configuration. Hence, it becomes conservative to make any segmentations but only deploys given commodities. This experiment only shows influences caused by quality of data sets. We keep the parameter configuration from *Experiment 1*. In practice, we should reconfigure system parameters while dealing with different data sets.

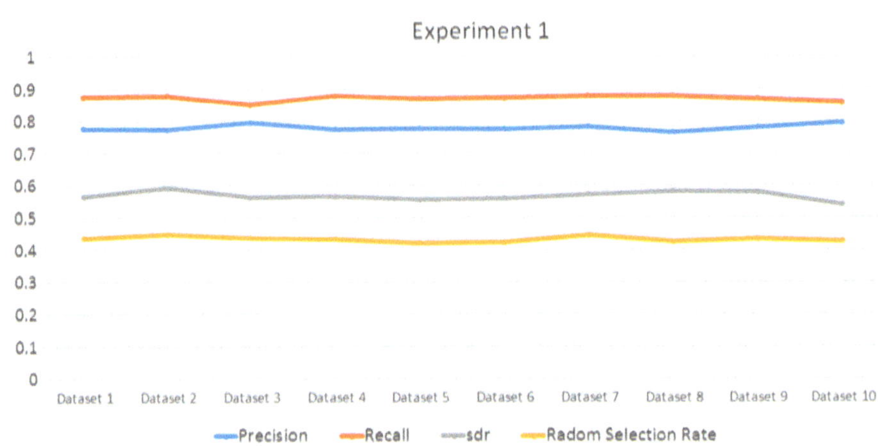

Fig. 1.1 Results of 10 disjoint test data sets with 20 % hidden links

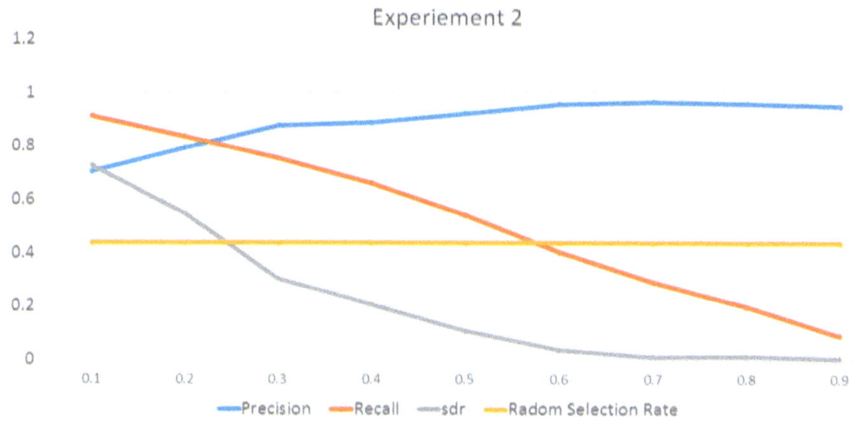

Fig. 1.2 Results of 10 test data sets with varying hidden links

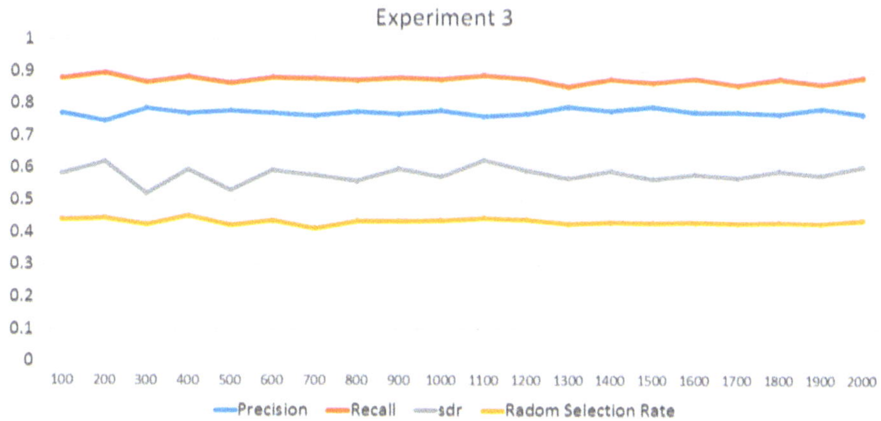

Fig. 1.3 Results of 20 test data sets with different size

In *Experiment 3*, we generate 20 test data sets with size varying from 100 to 2000. We keep the same configuration as in *Experiment 1*. The results are in Fig. 1.3. As shown from the figures, the performance of our approach isn't affected much by the size of the data sets. In *Experiment 3*, fluctuation of sdr is stronger than other metrics. However, by increasing the data set size, sdr becomes flatter. Prediction of sdr is hard. As shown in *Experiment 2*, a data set should not include many unknown information otherwise segmentations will be inaccurate. Correlating segmentations in a dual network becomes harder when one side of the dual network has much less segments than the other side.

In *Experiment 4*, we compare our solution with some other existing approaches for market segmentation over the same data sets. The problem is that some approaches

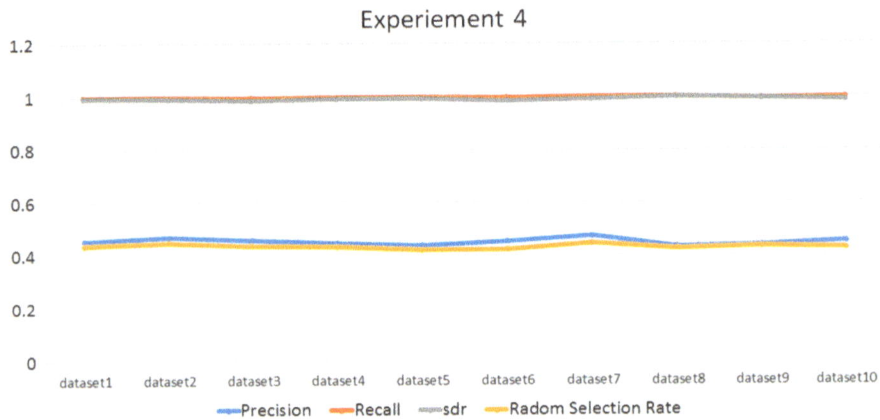

Fig. 1.4 Computing the results of our PSA-based approach with other approaches

such as k-means, x-means and hierarchical clustering require predetermined numbers of segmentations, which different from our approach where no prior knowledge to the number of segments is needed. Therefore we only compare our solution to algorithms that can compute with an unknown number of segments. The expectation maximization algorithm in WEKA uses cross validation to determine the number segmentations in a data set. Our PSA system determines the number of segmentations on acceptances of different proposals. The results of *Experiment 4* are given in Fig. 1.4. EM clustering has high recall and sdr because it deploys all commodities to every consumer segmentation. It is only slightly better if a system randomly guesses the result. The advantage of PSA system is to dynamically segmenting consumers according to commodity segmentation. A proposal captures local information regarding to what requires in a segmentation.

Due to the mentioned inbalanceness of Jester data set, we further validate our approach in a balanced network where the number of consumers roughly equals to the number of commodities. To make the synthetic data more realistic, we includes certain noise data, so that certain consumers or commodities in a designated segment have an usual low number of links. Such noise does not affect the experimental results as our system still finds more hidden segments while keeping a high value of precision and recall. Our approach achieves an overall good performance on this synthesised data:

Precision	Recall	sdr	Random selection rate
0.87466	0.939716	0.981968	0.655504

1.5 Related Works

Huang et al. implements a market segmentation mechanism based on support vector machines (SVM) [6]. The authors conduct experiments on a drink company data set and demonstrate how SVM may outperform in terms of forming better clusterings than other techniques such as the k-means algorithm. In [14], Wang adopts another clustering techinque, namely robust fuzzy clustering. He improves the original algorithm by pre-processing data set so that noise data can be eliminated. This results in certain improvements in terms of quality of clusters. Migueis et al. use a variable clustering algorithm to determine consumer segmentations from purchase history dataset in [11]. The authors assign consumers to each segments to illustrate purchasing power of customers. Deng et al. analyse market segmentation in mobile e-commerce of a Japanese chain restaurant in [3]. They propose a hybrid clustering framework that combines k-means algorithm, self-organising feature mapping and particle swarm optimisation. They identified over 70 % observed customer segmentations; and each individual approach in the hybrid approach performs less accurate. Kuo et al. investigate different combinations of a set of algorithms in [7]. The authors conduct experiments on data collected from survey about web user segmentations and evaluate their results through a business marketing analysis.

1.6 Conclusion and Future Work

This paper proposes a new approach that is based on the PSA-framework to simulate and solve the dual market segmentation problem. Our experimental results shows that on real data sets our approach may find reasonable segments while improving from existing approaches. Furthermore, as the approach is implemented in a distributed framework, it may be used on markets where data is maintained dynamically and detect the changing patterns of market segments.

As future works, there are several possible ways to extend our current work:

- Firstly, one could explore different models for implementing the PSA-system in our approach. In particular, one could use different similarity functions, utility functions and proposal functions and test their impact on the performance of the method.
- Secondly, more tests could be performed on real-world and artificial data. In particular, we would need data sets that contains a rather sophisticated market that contains a large number of consumers and a large number of products of different ranges. Instead of using data mining techniques such as cross validation, we could compare the identified market segments with real world ground-truth.
- Thirdly, the challenging aspect for our current approach is that we have to manually configure parameters such as different threshold in our method. To facilitate real automated market segmentation, it would be necessary to implement a mechanism that allows automated parameters configuration to best fit the data sets.

References

1. Campbell, C.: I shop therefore i know that i am: the metaphysical foundations of modern consumerism. In: Ekstrom, K., Brembeck, H. (eds.) Elusive Consumption, pp. 10–21. Berg, Oxford (2004)
2. Davis, J., Goadrich, M.: The relationship between Precision-Recall and ROC curves. In: Proceedings of the 23rd International Conference on Machine Learning, pp. 233–240. ACM (2006)
3. Deng, X.Y., Jin, C., Higuchi, Y., Han, J.C.: An efficient hybrid clustering algorithm for consumer segmentation in mobile e-commerce. ICIC Express Lett. **5**(4B), 1411–1416 (2011)
4. Dutta, S., Bhattacharya, S., Guin, K.K.: Data mining in market segmentation: a literature review and suggestions. In: Proceedings of Fourth International Conference on Soft Computing for Problem Solving, pp. 87–98. Springer, India
5. Goldberg, K., Roeder, T., Gupta, D., Perkins, C.: Eigentaste: a constant time collaborative filtering algorithm. Inf. Retrieval **4**(2), 133–151 (2001)
6. Huang, J.J., Tzeng, G.H., Ong, C.S.: Marketing segmentation using support vector clustering. Expert Syst. Appl. **32**(2), 313–317 (2007)
7. Kuo, R.J., Chang, K., Chien, S.Y.: Integration of self-organizing feature maps and genetic-algorithm-based clustering method for market segmentation. J. Organ. Comput. Electron. Comm. **14**(1), 43–60 (2004)
8. Li, Z., Wei, Z., Jia, W., Sun, M.: Daily life event segmentation for lifestyle evaluation based on multi-sensor data recorded by a wearable device. In: 2013 35th Annual International Conference of the IEEE Engineering in Medicine and Biology Society (EMBC), pp. 2858–2861. IEEE (2013)
9. Liu, J., Wei, Z.: Community detection based on graph dynamical systems with asynchronous runs. In: 2014 Second International Symposium on Computing and Networking (CANDAR), pp. 463–469. IEEE (2014)
10. Liu, J., Wei, Z.: From a local to a global perspective of community detection in networks. In: PRICAI 2014: Trends in Artificial Intelligence, pp. 1036–1049. Springer International Publishing (2014)
11. Miguéis, V.L., Camanho, A.S., e Cunha, J. F.: Consumer data mining for lifestyle segmentation. Expert Syst. Appl. **39**(10), 9359–9366 (2012)
12. Miguéis, V.L., Camanho, A.S., e Cunha, J.F.: Mining consumer loyalty card programs: The improvement of service levels enabled by innovative segmentation and promotions design. In: Exploring Services Science, pp. 83–97. Springer, Berlin (2011)
13. Tuner, J.C.: Redicovering the Social Group: A Self-categorization Theory, pp. 42–67. Blackwell, Oxford (1987)
14. Wang, C.H.: Apply robust segmentation to the service industry using kernel induced fuzzy clustering techniques. Expert Syst. Appl. **37**(12), 8395–8400 (2010)

Chapter 2
A Multiagent-Based Domain Transportation Approach for Optimal Resource Allocation in Emergency Management

Jihang Zhang, Minjie Zhang, Fenghui Ren and Jiakun Liu

Abstract In metropolitan regions, emergency events request urgent response within a short time limit in order to minimise the damage and the number of fatality. Most of these events require different resources that are usually distributed over a large area. How to efficiently allocate the distributed resources to an event is a challenging research issue. Traditional centralised resource allocation approaches have difficulties to find out the best resource assignment within the event's time limits by considering the dynamics of the metropolitan environment and the event itself. In this paper, a multiagent-based decentralised resource allocation approach using domain transportation theory is proposed to handle an emergency event with multiple tasks. Experimental results indicates that the proposed approach can effectively generate the optimal resource allocation plans by considering multiple factors of an emergency event.

Keywords Domain transportation · Resource allocation · Emergency management

J. Zhang (✉) · M. Zhang · F. Ren
School of Computing and Information Technology,
University of Wollongong, Wollongong, Australia
e-mail: jz718@uowmail.edu.au

M. Zhang
e-mail: minjie@uow.edu.au

F. Ren
e-mail: fren@uow.edu.au

J. Liu
School of Mathematics and Applied Statistics,
University of Wollongong, Wollongong, Australia
e-mail: jiakunl@uow.edu.au

Q. Bai et al. (eds.), *Multi-agent and Complex Systems*, Studies in Computational Intelligence 670, DOI 10.1007/978-981-10-2564-8_2

2.1 Introduction

In recent decades, with the rapid increase of population in metropolitan regions, public emergency departments suffer significant pressure about how to efficiently and effectively allocate rescue resources for emergency events (i.e. vehicle accident, fire, terrorist attacks, etc.). Generally, these emergency events have the three common characteristics, including that (1) they are hard to be predicted in dynamic environments such as metropolitan regions; (2) they usually require multiple different resources with different usage costs, mobilities, availabilities, ownerships and functionalities (i.e. a vehicle accident might require ambulances, polices and fire fighters); and (3) they have strict time limits for emergency departments to response and allocate rescue resources. Currently, most resource allocation processes for the emergency events in metropolitan regions are still operated manually, which is highly inefficient. This is because in a metropolitan region, large numbers of rescue resources with different functionalities and availabilities are distributed over an extensive area. Emergency event operators usually have difficulties to efficiently find out the optimal resource allocation for an emergency event due to a large number of possibilities.

In recent years, agent and multi-agent systems have been becoming promising technologies for resource allocation in emergency situations and many useful approaches have been proposed to assist human operators to make decisions [1–3, 5, 6]. Aygul et al. [3] proposed an agent-based solution for finding the optimal allocation of emergency medical resources, which was implemented based on service-oriented architecture. Since their approach requires the global information of all resources, so it might not available in most real life situations. Beatriz et al. [5] proposed a multi-agent system to allocate ambulances for emergency medical events by using the contract net protocol and a winner determination algorithm to find out the optimal ambulances allocation. However, their approach is restricted to handle ambulances, so it might not suitable for emergency events with the requirements of different resources. Fiedrich [1] proposed an multi-agent system for optimal resource allocation for large natural disasters based on High Level Architecture (HLA), which was capable of sequentially allocating appropriate resources to tasks. However, sequential resource allocation might increase the number of fatality of tasks in the later position of the task queue.

In order to overcome the above limitations, this paper proposes an agent-based decentralised resource allocation approach for an emergency event in metropolitan regions. The proposed approach first converts the resource allocation problem of a single emergency event into different resource allocation tasks. Then, different agents propose their resource allocation proposals to these tasks simultaneously. Finally, for each task, the domain transportation theory is used to combines these proposals and find out the optimal resource allocation that minimises the total allocation cost. The major contributions of the proposed approach are that (**1**) the proposed approach is designed to efficiently allocate distributed resources in decentralised manner, which is more applicable and practical comparing with centralised approaches; (**2**) the proposed approach is designed to effectively handle different types of emergence events that require resources with different functionalities; (**3**) the proposed resource allocation framework can simultaneously allocate different resources to different tasks of an emergency event; and (**4**) the proposed resource allocation algorithm is highly adaptable and flexible, which is capable of generating optimal allocation solutions for emergency situations in different domains with different cost functions or attributes.

The rest of this paper is organised as follow. Section 2.2 gives problem description and definitions. Section 2.3 introduces the theoretical foundation of the optimal resource allocation. Section 2.4 describes the agent-based solution for the implementation of the proposed resource allocation approach. Section 2.5 demonstrates the experimental results and provides analysis. Section 2.6 gives the conclusion and outlines the future work.

2.2 Problem Description and Definitions

This section introduces the important definitions that are used in the proposed resource allocation approach and the fundamental problem that the proposed approach is trying to address.

2.2.1 Definitions of Domain Knowledge

Definition 1 (*Environment*) An *environment* is represented by a city map, which is defined as an undirected graph, $G = [\mathbb{V}, \mathbb{E}]$, where $\mathbb{V} = \{v_1, v_2, ..., v_i\}$ is a set of nodes, which represent important locations in a metropolitan region, and $\mathbb{E} = \{e_1, e_2, ..., e_j\}$ is a set of edges, which represent the paths between the nodes. e_j is further defined as a two-tuple, $e_j = (v_o, v_p)$, where $v_o, v_p \in \mathbb{V}$ are the nodes that be connected by e_j.

Definition 2 (*Resource*) A *resource* is defined as a seven-tuple, $res = (rty, ser, fun, rlo, ava, vel, exp)$, where $rty \in \{facility, mobile\}$ represents the type of resources, where facility refers to unmovable rescue resources, such as fire stations, hospitals, etc., and mobile refers to rescue vehicles and personnel; $ser \in$ {*fire & rescue, medical, police*} represents the type of emergency services that res can provide; $fun \in$ {*fire fighter, fire truck, police officer, police car, ambulance, medical personnel*} represents res 's functionality; $rlo \in \mathbb{V}$ represents res 's current location; $ava = \{0, 1\}$ represents res 's availability, where 0 indicates unavailable and 1 indicates available; $vel \in (0, +\infty)$ represents res 's average velocity in kilometre per hour (km/h) and exp represents res 's money expenditure in per hour when $rty = facility$, while exp represents res 's money expenditure in per hour, per kilometre when $rty = mobile$.

In the proposed approach, set $\mathbb{REE}$ indicates all resources in G. Besides, it is assumed that the money expenditure exp of res is known by local emergency departments. Furthermore, the set of resource services and functionalities defined above could be extended in real-world applications.

Definition 3 (*Task*) A *task* is defined as a three-tuple, $tas = (dea, ser, \mathbb{TR})$, where dea represents the deadline for resources to be allocated to tas; ser represents the type of emergency service that tas requires to make response and $\mathbb{TR} = \{tr_1, tr_2, ..., tr_e\}$ represents a set of required resources for completing tas. tr_e is further defined as a two-tuple, $tr_e = \{rty, fun\}$.

In the proposed approach, it is assumed that local emergency departments have the knowledge to estimated a task's deadline dea based on an event's severity.

Definition 4 (*Event*) An *event* is defined as a five-tuple, $eve = (con, \mathbb{SER}, elo, sev, \mathbb{TAS})$, where $con \in$ {*fire, rescue, loss of life, damage to health, security of person, security of property*} represents eve 's content; $\mathbb{SER} \subseteq$ {fire & rescue, medical, police} represents a set of emergency services required by eve; $elo \in \mathbb{V}$ represents eve 's location; $sev \in [1, 10]$ represents eve 's severity and $\mathbb{TAS} = \{tas_1, tas_2, ..., tas_k\}$ represents a sequence of tasks that need to be completed for eve.

Definition 5 (*Resource Allocation Proposal*) A *resource allocation proposal* for an event is defined as a two-tuple, $rap = (eve, \mathbb{RES})$, where $\mathbb{RES} \subseteq \mathbb{REE}$ represents a set of resources that be proposed for completing tasks in eve. Besides, a *resource allocation proposal* for a single task in the event is defined as a two-tuple, $rap_k = (tas_k, \mathbb{RES})$, where $rap_k.\mathbb{RES} \subseteq rap.\mathbb{RES}$.

In the proposed approach, the cost of resource allocation is calculated by cost functions. Usually, different emergency events might need to use different cost functions, which might involve different cost attributes. At here, a cost function is defined, by considering two significant factors in emergency resource allocation, i.e., money expenditure and time. In the following, the cost for allocating a single resource to a single task is firstly defined.

Definition 6 (*Cost Function*) A *cost function* for a single resource's allocation is defined by Eq. 2.1:

$$COR(eve.tas_k, res) = \begin{cases} res.exp, \text{ if } res.rty = facility \\ res.exp \times DIS(res.rlo, eve.elo) \times \frac{DIS(res.rlo, eve.elo)}{res.vel} \times \\ \qquad DLINE(eve.tas_k, res), \text{ if } res.rty = mobile, \end{cases} \tag{2.1}$$

where $DIS(res.rlo, eve.elo)$ is a function that return the distance of a passable road between resource location $res.rlo$ and event location $eve.elo$, which could be implemented by various path searching algorithms, such as Dijkstra's and A* algorithms [7, 9]. $DLINE(eve.tas_k, res)$ is a function that is used to determine whether res can be allocated within $eve.tas_k$'s deadline, which is further defined by Eq. 2.2:

$$DLINE(eve.tas_k, res) = \begin{cases} 1, \text{ if } \frac{DIS(res.rlo, eve.elo)}{res.vel} \leq eve.tas_k.dea \\ +\infty, \text{ if } \frac{DIS(res.rlo, eve.elo)}{res.vel} > eve.tas_k.dea \end{cases} \quad (2.2)$$

Based on above terms, the cost function for allocating all required resources to a single task is defined by Eq. 2.3:

$$COT(eve.tas_k, rap_k) = \sum_{res_v \in rap_k.\mathbb{RES}} COR(eve.tas_k, res_v) \quad (2.3)$$

Furthermore, the cost function for allocating all required resources to all tasks in a single event is defined by Eq. 2.4:

$$COE(eve, rap) = \sum_{eve.tas_k \in eve.\mathbb{TAS}} COT(eve.tas_k, rap_k) \quad (2.4)$$

2.2.2 Problem Description

For an emergency event eve, there could be different resource allocation proposals. In the proposed approach, the all possible resource allocation proposals for eve are represented as set $\mathbb{RAP}$. The main objective of the proposed approach is to search an optimal resource allocation proposal $rap^* \in \mathbb{RAP}$ for eve. The objective function for an event's resource allocation is formally defined by Eq. 2.5:

$$OBJE = \arg \min_{rap^* \in \mathbb{RAP}} COE(eve, rap^*) \text{ subject to } rap^* \in \mathbb{REE} \quad (2.5)$$

where $OBJE$ represents the name of objective function for an event's resource allocation and $rap^* \in \mathbb{REE}$ means that the proposed resources in $rap^*.\mathbb{RES}$ must belong to the available resources in the environment G. The event objective function indicates that the optimal resource allocation proposal rap^* must have the minimal allocation cost in $\mathbb{RAP}$. Besides, the proposed approach assumes there is always enough resources in $\mathbb{REE}$ to be allocated for eve.

However, due to the fact that an emergency event usually require resources with different types and functionalities, searching a complete rap^* could be a complicated and time-consuming process. In order to efficiently solve this searching problem, the proposed approach creates a set of resource allocation tasks $eve.\mathbb{TAS}$ for eve. For each task tas_k in $eve.\mathbb{TAS}$, it only requires resources that provide the same type of emergency service (i.e. $res.ser$). By doing so, the searching of rap^* for eve is converted to the concurrent searching of $rap_k^* \in \mathbb{RAP}_k$ for each task tas_k in $eve.\mathbb{TAS}$, where $\mathbb{RAP}_k$ represents the all possible or available resource allocation proposals for tas_k. The objective function of resource allocation for a task tas_k is formally defined by Eq. 2.6:

$$OBJT = \arg \min_{rap_k^* \in \mathbb{RAP}_k} COT(eve.tas_k, rap_k^*) \text{ subject to } rap_k^*.\mathbb{RES} \in \mathbb{REE} \quad (2.6)$$

where $OBJT$ represents the name of the objective function for an task's resource allocation.

2.3 Theoretical Foundation of the Domain Transportation for the Optimal Resource Allocation

Domain transportation theory is a linear programming method to minimise the cost of relocating resources [4]. In domain transportation, the resource allocation problem of a task can be described as a resource mapping problem from the available resources in an environment to the required resources of the task, which is formally represented by Eq. 2.7:

$$rap_k : \mathbb{REE} \rightarrow tas_k.\mathbb{TR} \quad (2.7)$$

Apparently, there are many different mapping proposals (i.e. $\mathbb{RAP}_k$) from domain $\mathbb{REE}$ to domain $tas_k.\mathbb{TR}$. In this paper, domain transportation theory is used to find out rap_k^* for tas_k, which can fulfil Eq. 2.6.

More precisely, let $\mathbb{REE}_e(y_i)$ denote the amount of functionality e resource at location $y_i \in \mathbb{V}$. Let $x_k = eve.tas_k$ and $tr_e(x_k)$ represent the required amount of functionality e resource at task x_k. The cost of transferring the resource at y_j to the task x_k is given by Eq. 2.1, i.e. $COR(x_k, y_j)$. An admissible allocation proposal rap_k is a mapping from $\mathbb{V}$ to $\mathbb{TAS}$ satisfying the balance condition that for any subset $E \subset \mathbb{TAS}$,

$$\sum_{x_k \in E} tr_e(x_k) = \sum_{y_j \in rap_k^{-1}(E)} \mathbb{REE}_e(y_j), \quad (2.8)$$

where rap_k^{-1} is the inverse mapping of rap_k. As before, let $\mathbb{RAP}_k$ denote the set of all admissible allocation proposals. The purpose of Eq. 2.6 is to find an optimal proposal $rap_k^* \in \mathbb{RAP}_k$ such that

$$COT(x_k, rap_k^*) = \min_{rap_k \in \mathbb{RAP}_k} COT(x_k, rap_k). \quad (2.9)$$

From optimal transport theory, Eq. 2.9 can be transferred to the following linear programming

$$\max \left\{ \sum_{x_k \in \mathbb{TAS}, y_j \in \mathbb{V}} u(x_k)tr_e(x_k) + v(y_j)\mathbb{REE}_e(y_j) : u(x_k) + v(y_j) \le COR(x_k, y_j) \right\}. \tag{2.10}$$

Moreover, there exists a maximiser (u^*, v^*) (unique up to a constant) achieving the above maximum. u^* is called a potential, and v^* is its dual potential. The pair (u^*, v^*) also satisfies a generalised Legendre duality associated with the cost function COR. Hence, for each x_k, there exists a unique y_i such that the equality holds in the constraint, namely $u^*(x_k) + v^*(y_i) = COR(x_k, y_i)$, which means the task x_k requires the resource from location y_i. Thus, one can construct a mapping $inv_k^* : x_k \in \mathbb{TAS} \to y_j \in \mathbb{V}$. From optimal transport theory, by differentiating the above equation one can see that

$$\begin{aligned} \nabla u^*(x_k) &= \nabla_x COR(x_k, inv_k^*(x_k)), \\ y_i &= inv_k^*(x_k) = \nabla_x^{-1} COR(x_k, \nabla u^*(x_k)). \end{aligned} \tag{2.11}$$

In fact, one can further show that inv_k^* is exactly the inverse of the optimal allocation proposal rap_k^*.

Therefore, to construct rap_k^*, it suffices to follow the following steps:

1. From the given data $\{tr_e, \mathbb{TAS}, \mathbb{REE}_k, \mathbb{V}\}$, formulate the linear programming Eq. 2.10;
2. Solve Eq. 2.10 to find out a potential u^*;
3. Using Eq. 2.11 to construct the mapping $inv_k^* : x_k \in \mathbb{TAS} \to \mathbb{V}$, which implies that task x_k requires the resource from the area $REQ := inv_k^*(x_k) \subset \mathbb{V}$;
4. Take the inverse, we obtain the optimal allocation proposal $rap_k^* : REQ \to x_k$, which can inform the agent how to distribute the resource in the optimal way.

2.4 Agent-Based Decentralised Resource Allocation

The proposed resource allocation approach is implemented based on agent and multi-agent technologies, due to agents ability of autonomous reasoning, intelligent decision making, group coordination and collaboration [8]. This section gives detail description of agents' definitions, resource allocation framework and process.

2.4.1 Definitions of Agents

Generally, there are four types of agents in the proposed resource allocation approach, which are ***response agent, mobile agent, facility agent*** and ***deployment agent***. Each agent's definition is described as follows.

Definition 7 (*Response Agent*) A *response agent* is represented by ra, which has the information of a specific emergency event. A response agent has four major functionalities, which are (1) identifying event content $eve.con$ for a new event eve; (2) identifying the emergency services $eve.\mathbb{SER}$ that is required by eve based on $eve.con$; (3) identifying a set of tasks $eve.\mathbb{TAS}$ for eve based on $eve.ser$; and (4) sending $eve.\mathbb{TAS}$ to a deployment agent.

Definition 8 (*Mobile Agent*) A *mobile agent* is represented by ma, which has the information of a specific mobile resource $ma.res$. A mobile agent has two major functionalities, which are (1) managing a mobile resource $ma.res$; and (2) implementing resource allocation paths after receiving resources allocation commands.

Definition 9 (*Facility Agent*) A *facility agent* is represented by fa, which has the information of a specific facility resource $fa.res$ and a set of mobile agents. More precisely, $fa.\mathbb{MA} = \{ma_1, ma_2, ..., ma_j\}$ represents a set of mobile agents that belong to fa and $\mathbb{REF} = \{fa.res\} \cup \{ma_1.res, ma_2.res, ..., ma_j.res | ma_j \in fa.\mathbb{MA}\}$ represents all resources under fa 's management. A facility agent has three major functionalities, which are (1) managing a facility resource $fa.res$; (2) generating resource allocation proposals for tasks based on $fa.\mathbb{REF}$; and (3) informing its mobile agents to execute resources allocation commands after receiving the commands from a deployment agent.

Definition 10 (*Deployment Agent*) A *deployment agent* is represented by da, which has the information of a specific emergency event. A deployment agent has three major functionalities, which are (1) informing an event's tasks information (i.e. $eve.\mathbb{TAS}$) to facility agents that are located in its circle communication area, represented by $da.com$; (2) combining and generating the optimal resource allocation proposal rap_k^* for a task based on a set of proposals (i.e. $\mathbb{RAP}_k$) submitted by facility agents; and (3) informing relevant facility agents to execute rap_k^*.

2.4.2 Resource Allocation Framework and Process

The proposed resource allocation approach is implemented by a multi-agent system (MAS), which includes a task identification module, resource identification module, proposal generation module, optimal allocation module and proposal execution module. The framework of the MAS is depicted in Fig. 2.1.

As depicted in Fig. 2.1, an event's resource allocation process involves one response agent, one deployment agent, multiple mobile agents and facility agents. The allocation process is formally described by Algorithm 1.

Algorithm 1 : Resource Allocation Process

1: assign ra to eve
2: ra identifies $eve.con$
3: ra identifies $eve.\mathbb{SER}$ based on $eve.con$
4: ra identifies $eve.\mathbb{TAS}$ based on $eve.\mathbb{SER}$
5: ra sends $eve.\mathbb{TAS}$ to da
6: da calculates circle communication area $da.com$
7: da locates $\mathbb{FA}$ in $da.com$
8: **for all** $tas_k \in eve.\mathbb{TAS}$ **do**
9:     da creates $\mathbb{RAP}_k$ and $\mathbb{FA}_k$
10:     **for all** $fa_i \in \mathbb{FA}$ **do**
11:         **if** $fa_i.res.ser = tas_k.ser$ **then**
12:             da updates $\mathbb{FA}_k = \{fa_i\} \cup \mathbb{FA}_k$
13:             da sends tas_k to fa_i
14:             fa_i finds $rap_k^i : fa_i.\mathbb{REF} \rightarrow tas_k.\mathbb{TR}$
15:             **if** $tcu \leq PDLINE(tas_k.dea, eve.sev)$ **then**
16:                 fa_i submits rap_k^i to da
17:                 da updates $\mathbb{RED}_k = \{rap_i\} \cup \mathbb{RAP}_k$
18:             **end if**
19:         **end if**
20:     **end for**
21: **end for**
22: **while** $|eve.\mathbb{TAS}| > 0$ **do**
23:     **for all** $tas_k \in eve.\mathbb{TAS}$ **do**
24:         **if** $tcu \geq PDLINE(tas_k.dea, eve.sev) \vee \forall fa_g \in \mathbb{FA}_k : fa_g$ submit rap_k^g **then**
25:             **if** $\mathbb{RAP}_k$ contains enough resources for tas_k **then**
26:                 da expend $da.com$ by double
27:                 process goes back to Line 7
28:             **else**
29:                 da finds $rap_k^* : \mathbb{RED}_k \rightarrow tas_k.\mathbb{TR}$
30:                 da updates $eve.\mathbb{TAS} = eve.\mathbb{TAS} \setminus \{tas_k\}$
31:                 da informs agents to execute rap_k^*
32:             **end if**
33:         **end if**
34:     **end for**
35: **end while**

The resource allocation process shown in Algorithm 1 includes six steps, which are explained as follows.

Step 1: (Lines 1–5) When an emergency event eve happened, a new response agent ra is assigned to eve to identify the emergency content $eve.con$. Then, ra needs to identify the emergency services $eve.\mathbb{SER}$ required by eve according to $eve.con$. For example, when $eve.con = fire$, the required emergency services could be $eve.\mathbb{SER}$ = *{fire & rescue, medical, police}*. After emergency service identification, ra needs to acquire the information of the resources required by each emergency service, which may provided by human operators or other external agents. Then, ra converts each of emergency service to a task. Finally, ra sends $eve.\mathbb{TAS}$ to a deployment agent da.

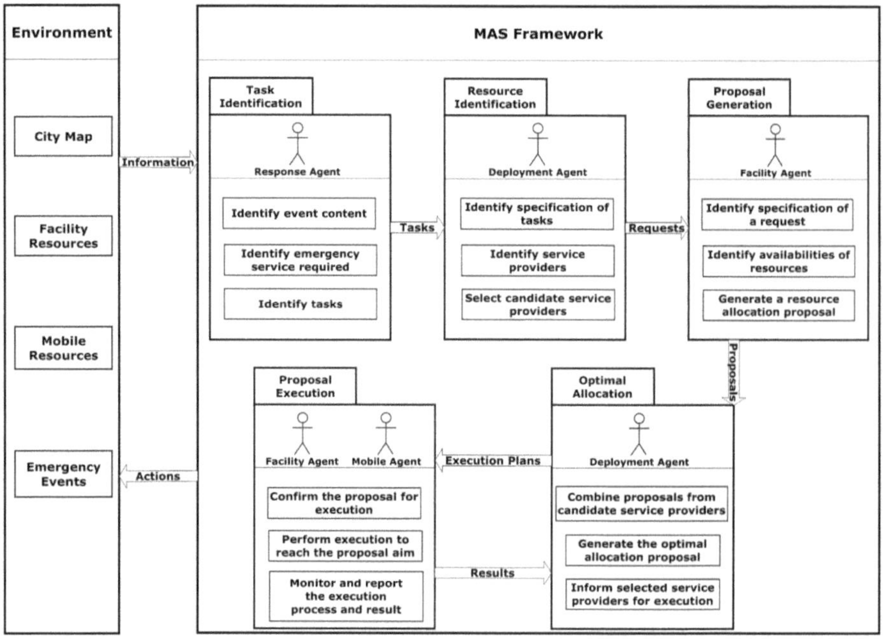

Fig. 2.1 The framework of the proposed resource allocation system

Step 2: (Lines 6–13) After receiving $eve.\mathbb{TAS}$, da first needs to calculate an communication area $da.com$, which is a circle centred at the event's location $eve.elo$ and measured by square kilometres. $da.com$ is calculated by Eq. 2.12:

$$da.com = \pi \times (avv \times (\frac{\sum_{task \in eve.\mathbb{TAS}} task_k.dea}{|eve.\mathbb{TAS}|} - tcu))^2 \qquad (2.12)$$

where avv represents the average moving velocity (km/h) of all required mobile resources in $eve.\mathbb{TAS}$ and tcu represents the current time. In the proposed approach, it is assumed that local emergency departments have the knowledge of the average velocity of different functionalities' resources.

After the calculation of $da.com$, da needs to locate all facility agents inside $da.com$, which is represented by set $\mathbb{FA} = \{fa_1, fa_2, ..., fa_i\}$ (Line 9). Then, da sends each tas_k to relevant facility agents in $\mathbb{FA}$ based on tas_k 's emergency service requirement $tas_k.ser$. At the same time, da also creates a facility agent contact list $\mathbb{FA}_k = \{fa_1, fa_2, ..., fa_g\}$ ($\mathbb{FA}_k \subseteq \mathbb{FA}$) and resource allocation proposal list $\mathbb{RAP}_k = \{rap_k^1, rap_k^2, ..., rap_k^g\}$ for each tas_k in $eve.\mathbb{TAS}$.

Step 3: (Lines 14–17) After a facility agent $fa_i \in \mathbb{FA}$ receives tas_k, fa_i uses the domain transportation theory (see Sect. 2.3) to calculate an optimal resource allocation proposal rap_k^i based on all available resources that under fa_i 's management (Line 14). After the calculation of rap_k^i, fa_i submits rap_k^i to da if current time tcu has not exceeded task proposal deadline. The task proposal deadline is calculated by function $PDLINE(tas_k.dea, eve.sev)$, which can be defined by local emergency departments based on the detail of tas_k and eve. After the submission of rap_k^i, da adds rap_k^i to $\mathbb{RAP}_k$.

Step 4: (Lines 23–25) After da receives tas_k 's resource allocation proposals from all facility agents in $\mathbb{FA}_k$ or tas_k 's proposal deadline has been reached, da checks whether $\mathbb{RAP}_k$ has enough resources for da to generate an final resource allocation plan to complete tas_k. If the resources are enough, the process goes to Step 6. Otherwise, the process goes to Step 5.

Step 5: (Lines 26–27) If $\mathbb{RAP}_k$ does not have enough resources to complete tas_k, da expands its original communication area $da.com$ by double to contact more facility agents, and then the process goes back to Step 2.

Step 6: (Lines 29–31) If $\mathbb{RAP}_k$ has enough resources to complete tas_k, da uses domain transportation theory (see Sect. 2.3) to generate an optimal resource allocation proposal rap_k^* for tas_k based on all resources in $\mathbb{RAP}_k$, which is represented by $\mathbb{RED}_k$ (Line 24). Finally, da informs relevant facility agents to execute rap_k^* and remove tas_k from $eve.\mathbb{TAS}$. If there are more tasks in $eve.\mathbb{TAS}$, the process repeats Step 4, otherwise the process ends.

2.5 Experiment

In this section, experimental results are presented and the performance of the proposed resource allocation approach is analysed. The experiments focus primarily on testing the resource allocation time, money expenditure and cost of an event when employing the proposed optimal resource mapping algorithm. The benchmark of the experiments is the decentralised ambulances allocation approach proposed by Beatriz et al. [5], which uses an auction mechanism based on trust and time to assign ambulances to emergency events.

2.5.1 Experimental Setting

In the experiments, the proposed resource allocation was tested in two different scenarios, which are: (**1**) resource allocation for an event with the proposed optimal resource allocation approach and (**2**) resource allocation for an event with Beatriz's ambulances allocation approach.

Although Beatriz's approach was specifically designed to allocate ambulances, but it is suitable to be used to allocate different kinds of mobile resources. Therefore, only mobile resources were required by the two scenarios' experiments. Furthermore, both two scenarios were conducted over an 1000×1000 grid. The experiments in each scenarios were repeated for 1000 times and the average resource allocation cost, time

Table 2.1 Parameters for resource's setting

rty	*rlo*	*ava*	*vel*	*exp*	*fun*
Mobile	[0, 0]–[1000, 1000]	{0,1}	[20–200]	[1–100]	Ambulance, fire truck, police car

Table 2.2 Parameters for event's setting

con	$\mathbb{SER}$	*elo*	*sev*	$\|\mathbb{TAS}\|$
Fire	Fire & rescue, medical, police	[500, 500]	5	3

Table 2.3 Parameters for task's setting

No.	*ser*	*dea* (min)	$\|\mathbb{TR}\|$	*tr.rty*	*tr.fun*
Task 1	Fire & rescue	30	15	Mobile	Fire truck
Task 2	Medical	40	10	Mobile	Ambulance
Task 3	Police	50	10	Mobile	Police car

and money expenditure of an event was recorded. The resource allocation cost was calculated by Eq. 2.1. For each time of the experiment, the resources' parameters were randomly regenerated based on Table 2.1, but the parameters of the event (Table 2.2) and its tasks (Table 2.3) remain same.

2.5.2 *Experimental Result and Analysis*

The experimental results are shown in Fig. 2.2. As we can see from Fig. 2.2a, Beatriz's approach (Scenario 2) requires slight less resource allocation time for an event comparing with our approach (Scenario 1). Nevertheless, from Fig. 2.2b, c, we can see

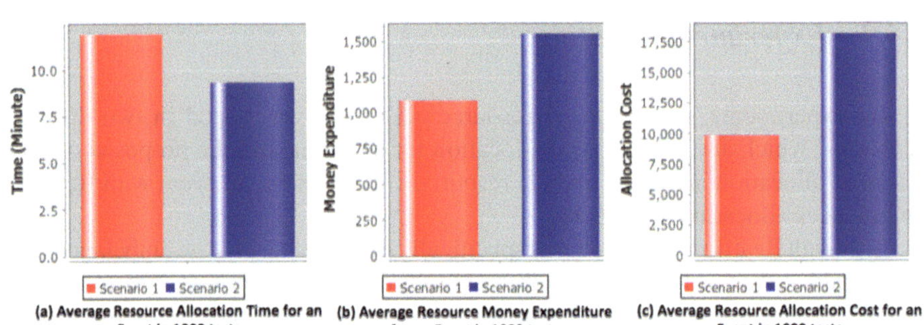

Fig. 2.2 Experimental results

that our approach outperforms Beatriz's approach significantly in terms of money expenditure (1086 vs. 1329) and allocation cost (9850 vs. 18100). This is mainly because in Beatriz's approach, resource allocation time is used as the primary criterion to determine the optimal allocation of resources, but our approach takes both resource allocation time and resource money expenditure into consideration. Another potential reason behind the results is that Beatriz's resource allocation approach was based on auction mechanism and the relationships between contract agents are competitive. However, for most emergency situations, it is more reasonable for agents to act cooperatively rather than competitively. In our approach, a deployment agent uses the domain transportation theory to generate the optimal solution based on the resource allocation proposals from multiple facility agents, so the optimal solution is a combined solution by integrating the advantages of each facility agent's proposal. By doing so, our approach can effectively minimise the required resource money expenditure and allocation cost by not compromising too much in terms of resource allocation time.

2.6 Conclusion and Future Work

In this paper, an agent-based decentralised resource allocation approach was proposed to handle an emergency event in metropolitan regions. In order to efficiently search an optimal resource allocation proposal for an event that requires different resources, the proposed approach first creates a set of tasks based on the emergency services required by the event. Then, domain transport theory is used to search the optimal resource allocation proposal for each task. The proposed approach was designed to handle multiple resource allocation tasks simultaneously and it can be used for emergency events in different domains. The experimental results indicate the good performance of the proposed approach in terms of resource allocation cost. Our future work will focus on handling concurrent emergency events and incorporating a resource coordination mechanism for resource contention problems between events.

References

1. Fiedrich, F.: An hla-based multiagent system for optimized resource allocation after strong earthquakes. In: Proceedings of the 38th Conference on Winter Simulation, pp. 486–492 (2006)
2. Fiedrich, F., Burghardt, P.: Agent-based systems for disaster management. Commun. ACM **50**(3), 41–42 (2007)
3. Gabdulkhakova, A., Konig-Ries, B., Rizvanov, D.A.: An agent-based solution to the resource allocation problem in emergency situations. In: 2011 Ninth IEEE European Conference on Web Services (ECOWS), pp. 151–157 (2011)
4. Liu, J.: Regularity of monge-ampère equations in optimal transportation. Bull. Austr. Math. Soc. **83**(01), 173–176 (2011)

5. López, B., Innocenti, B., Busquets, D.: A multiagent system for coordinating ambulances for emergency medical services. IEEE Intell. Syst. **23**(5), 50–57 (2008)
6. Martí, R., Robles, S., Martín-Campillo, A., Cucurull, J.: Providing early resource allocation during emergencies: The mobile triage tag. J. Netw. Comput. Appl. **32**(6), 1167–1182 (2009)
7. Skiena, S.: Dijkstra's algorithm. In: Implementing Discrete Mathematics: Combinatorics and Graph Theory with Mathematica, pp. 225–227 (1990)
8. Wooldridge, M., Jennings, N.R.: Intelligent agents: theory and practice. Knowl. Eng. Rev. **10**(02), 115–152 (1995)
9. Zeng, W., Church, R.L.: Finding shortest paths on real road networks: the case for a*. Int. J. Geogr. Inf. Sci. **23**(4), 531–543 (2009)

Chapter 3
CORPNET: Towards a Decision Support System for Organizational Network Analysis Using Multiplex Interpersonal Relations

Jiamou Liu, Anastasia Moskvina and Michael Ouředník

Abstract Organization network analysis (ONA) refers to a systematic exploration of organizational networks using techniques from social network theory, in the hope of better understanding the management structures, interpersonal relations and information flow within an organization. In this paper we introduce CORPNET, a standalone software application for ONA, which provides functions for simulating, analyzing and visualizing the interpersonal relations within corporations. This software tool is different from existing ONA tools in the sense that it is based on a multiplex network model which take into account both formal and informal relations between individuals in an organization. We demonstrate that CORPNET can compute individual influence utilizing the notions of network centrality, as well as providing useful guidelines regarding stability and leadership styles. Moreover, CORPNET also incorporates a novel benchmark network simulator which generates random informal links within a company. The overall goal is to develop an integrated intelligent decision support system based on organizational networks.

Keywords Organisational network analysis · Bonacich power · Formal and informal relations · Leadership styles

---

Availability: CORPNET and its source code can be downloaded from https://github.com/mourednik/corpnet

J. Liu (✉)
Department of Computer Science, The University of Auckland,
Auckland, New Zealand
e-mail: jiamou.liu@auckland.ac.nz

A. Moskvina
School of Engineering, Computer and Mathematical Sciences,
Auckland University of Technology, Auckland, New Zealand
e-mail: anastasia.moskvina@aut.ac.nz

M. Ouředník
e-mail: mike.ourednik@gmail.com

Q. Bai et al. (eds.), *Multi-agent and Complex Systems*, Studies in Computational Intelligence 670, DOI 10.1007/978-981-10-2564-8_3

3.1 Introduction

Rapid technological and industrial advancement in the last few decades facilitates expansion of companies into giant enterprises. As a company grows, it acquires a larger number of employees, henceforth encompasses increasingly complex interpersonal relations. To better understand the effectiveness and performance of the management structure, one should learn not only decisions made by top management, but also how information flows within the company through employee interactions. A major effort in management science is, therefore, to understand how the observed interpersonal relations affect the company performance as a whole. In particular, the emerging field of *organizational network analysis (ONA)* amounts to a systematic approach to analyze structures within a large organization, thereby providing useful decision supports [4, 5, 7].

We classify interactions among employees into two types: the *formal* structure, which represents the reporting hierarchy, and the *informal* structure, which represents the other, more personal connections, such as collaboration, friendship, and advices. As Krackhardt and Hanson pointed out in [11], if the formal structure is the skeleton of the company, the informal structure is its central nervous system. While the formal hierarchy defines positions in an organization, the informal structure largely influences information flow. While the formal structure is static, informal networks are more flexible and adaptive. Ignoring the informal structure could be a costly mistake since much of the work in a company happens despite the formal organization. Hence there is a need for an intelligent decision support system that combines techniques from both static and dynamic network analysis.

There are several existing software tools created for analysing networks within an organization. The reader is referred to, for examples, the softwares [10, 15, 16]. These software tools usually perform data visualization, as well as computing common network measures (centrality, network density, cluster analysis etc.). However, two significant limitations exist

1. Such products are mostly commercially available which made them difficult to be adopted for management science research, and,
2. More importantly, they do not study correlations between formal and informal structures with an organizational network.

In this paper, we introduce CORPNET, a stand-alone software application created to perform flexible organizational network analysis. The underlying organizational model adopted by CORPNET incorporates both formal and informal ties and hence is a multiplex network. The main features of CORPNET include instruments of statistical and stability analysis, community detection, and ability to generate real-life alike hierarchies and social networks. It also provides visualization of the network together with the key indicators above. Using a range of parameters, the software not only allows identification of personal influence in a company, but also reasoning about leadership styles and strategies.

Moreover, CORPNET is also able to generate synthetic benchmark networks, which could be used for ONA research. We point out that existing benchmark graphs in social network theory are not suitable case studies for ONA where employees in an organizations have a formal hierarchy as well as informal relations. Indeed, the interactions among people in an organization are often based on the *homophily principles*: people tend to make friends with those who are similar to them [13]. Therefore, informal networks in organizations generally reflect both the departments of individuals, and their levels of hierarchy. In this spirit, we propose a new model that takes into account hierarchical differences among employees. Thus, we enable generating and simulating real world organizational networks within our framework.

The rest of the paper is organized as follows. Section 3.2 introduces the multiplex network model for realising interpersonal relations in an organization. Section 3.3 discusses Bonacich power, which provides the foundation of individual influences in CORPNET. Section 3.4 presents main features of the CORPNET software. Section 3.5 discusses the benchmark networks generated by CORPNET. Section 3.6 demonstrates how CORPNET can be used to capture leadership styles and Sect. 3.7 concludes with remarks on future works.

3.2 Organizational Network Model of CORPNET

In this section we present the underlying network model of CORPNET. The reader is referred to [12] for a more thorough introduction to the theoretical models of interpersonal ties within an organizational structure. An organizational structure is often defined as a set of positions, groups of positions, reporting relationships and interaction patterns [1]. We use graphs as the basic structure to capture these notions: Nodes in the graph denote the work positions (or employees). There are two types of edges in the graph model:

1. An employee may report to a certain manager. We represent the reporting relation from any employee to her manager by an edge. If an employee does not report to anyone in the company, we refer to her as a *top-level manager* or a *root* (they could act, for example, as the board of directors). Moreover we also require that all top-level managers are mutually linked by edges. The edges created from the reporting relations and the relation between top-level managers are called *formal ties*, as they define the most official path of information inside the company. Clearly, the formal ties exhibit a strong association between people and are directed. Hence they form a directed graph.
2. The employees also are associated in a more "random" fashion thanks to different informal links. Such edges are referred to as *informal ties* as they may undergo frequent changes. Such informal ties are usually mutual between two people and represent a weaker association between people. Hence we assume the informal ties form an undirected graph.

On the one hand, the formal ties delineate the organizational hierarchy of a firm by featuring reporting relationship. On the other hand, the informal ties enrich the model by including informal relations among employees. More formally we define the following:

Definition 1 (*Organizational network*) An *organizational network* is a structure $\mathcal{G} = (V, R, E_s, E_w)$, where V is a set of *nodes*, $E_s, E_w \subseteq V^2$ are, respectively, the formal and informal tie relations with the following properties:

1. $R \subseteq V$ is a set of *roots* and $(r_1, r_2) \in E_s$ for any $r_1 \neq r_2 \in R$; When R only consists of a single node r (i.e., $|R| = 1$), there is an edge $(r, r) \in E_s$ (a self-loop).
2. the pair (V, E_s) forms a directed acyclic graph, where every node apart from nodes in R has an incoming edge from another node;
3. the pair (V, E_w) forms an undirected graph.

To refine this network model, we put forward constraints that reflect important principles in real-life organizational networks. In particular CORPNET follows the following principles:

1. **One Manager**. If a person has several sources of instructions, a head-on collision may occur which results in confusion and conflicts. Hence in an effective organization one would require every employee to report to at most one manager. Thus the formal tie relation on the subordinates below a certain root consists of a directed tree.
   Principle 1 (One Manager): Each node has exactly one incoming directed edge, which represents relationship with its manager, i.e., for all $u \in V$ there is a unique $v \in V$ with $(v, u) \in E_s$.

2. **Limited Capacity**. Capacity is an important concept in social network analysis which asserts that anyone can maintain only a limited number of interpersonal relations, due to limited time and effort [3]. For example, the number of strong personal links for networks of more than 500 nodes on Facebook varies from 10 to 20 [6].

   In our multiplex network model, we distinguish formal and informal ties in terms of how much resource each of them consumes. Let Δ be an abstract quantity that defines the maximum amount of resources (working hours, for instance) that a person can distribute between his or her ties. For simplicity, we assume that each person in the network has the same amount of resources Δ. We assume that a person needs s resources and w resources to maintain a formal and an informal tie, respectively. Therefore, for any node v, if $E_s(v)$ is the number of directed edges (including self-loop), and $E_w(v)$ is the number of undirected edges, then this person spends $E_s(v) \times s + E_w(v) \times w \leq \Delta$ resources to maintain all his connections.

Call the ratio $\delta = \frac{w}{s}$ the *correlation coefficient*. We define the *relative degree* of a node $v \in V$ as $d(v) = E_s(v) + E_w(v) \times \delta$, where $E_s(v)$, $E_w(v)$ are the numbers of formal ties (including both incoming and outgoing edges) and informal ties v maintains, respectively.
Principle 2 (Maximal Relative Capacity): There is a constant c such that for all $v \in V$, the relative degree $d(v) \leq c$.

3. **Limited Height** In management science, *flattening* or *delayering* refers to the process of elimination of layers in a firm's organizational hierarchy [21]. Back in 1950 s companies had up to twenty layers of hierarchy while by the end of the twentieth century they were trimmed to five or six. This phenomenon has been thoroughly investigated in the literature [19, 20]. Similarly to the "6-degree of separation" in social networks, in an effective organization, the top managers are usually not very far from all subordinates. Then, the third principle is simply as following:
Principle 3 (Limited Height): The largest level of hierarchy is not more than some given number h.

3.3 Individual Influences in an Organizational Network

A crucial functionality of CORPNET is the automatic identification of individual influences in an organization. Similarly to the concept of *social capitals* (see for example [6]), we define a notion of "positional advantage". Clearly, a position in a company is more advantageous if it is closer to the top managers. Moreover, a large number of personal connections provide more sources of information. Finally, the *span of control* indicates the number of subordinates to a manager, hence their level of involvement in making decisions within the network [9]. Based on this, we assign a *weight* between the values 0 and 1 to all edges e. The weight function μ depends on two parameters ρ and k:

- *Formal ties*: If $e \in E_s$, we assign $\mu_{\rho,k}(e) = 1$ if any end point of e is a non-root; and $\mu_{\rho,k}(e) = \rho$ otherwise. The edges between two roots (including the self-loop when there is only a single root) are special. A weight ρ of 1 suggests that it has the same affect as all the other formal ties, while 0 means that these edges only affects the capacity of the root node but not the influence. We will show later that the weight ρ is important for defining *management styles*: a larger ρ indicates a more "autocratic" style of management.
- *Informal ties*: If $e \in E_w$, we assign $\mu_{\rho,k}(e) = k$ where k is a parameter to the model. The range of k guarantees that directed edges are more important than undirected. An edge from A to B can be interpreted as A has influence over B ranked as the weight of this edge.
- *Back flow*: It is natural to assume that the influence between an employee and her manager is not one-way: while the manager influences the employee through a strong tie, the employee also influences her manager through social interaction, and

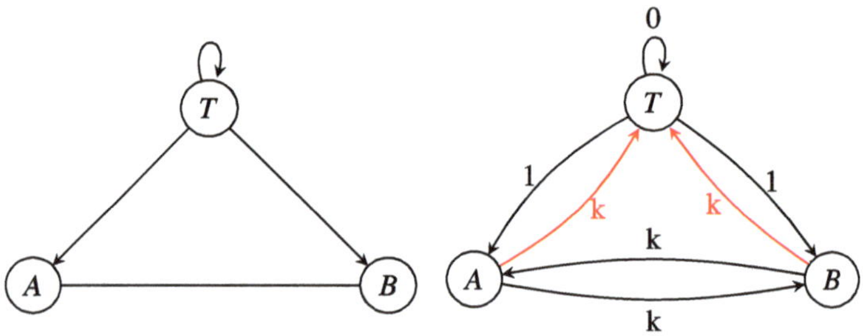

Fig. 3.1 An organizational network (*left*) and a weighted influence graph (*right*)

hence can be regarded as a informal tie, i.e., a back flow. Hence we set $\mu_{\rho,k}(e) = k$ where $e = (u, v)$ whenever u reports to v.

Based on the definition above, we regard any organizational network as a *weighted influence graph*; an example of this is shown in Fig. 3.1.

Definition 2 [*Weight influence graph*] Let $\mathcal{G} = (V, R, E_s, E_w)$ be an organizational network. The weighted influence graph of $\mathcal{G}$ is

$$W(\mathcal{G}) = (V, R, E_s, E_w, k, \rho, \mu_{\rho,k})$$

where k, ρ and the weight function $\mu_{\rho,k}$ are defined as above.

To capture individual influences in an organizational network, we use a centrality notion that is adapted from *Bonacich power*. Philip Bonacich in [2] expanded the eigenvalue centrality by introducing two additional parameters α and β. Let A be an adjacency matrix of a network. Then for any node i, the Bonacich power is defined as follows:

$$p_i(\alpha, \beta) = \sum_j (\alpha + \beta p_j) A_{ij} \tag{3.1}$$

We collect Bonacich power of all nodes in a vector $\mathbf{p}(\alpha, \beta)$:

$$\mathbf{p}(\alpha, \beta) = \alpha(I - \beta A)^{-1} A\mathbf{e}, \tag{3.2}$$

where I is an identity matrix, and $\mathbf{e}$ is a column vector of ones. The parameter α is a scalar that affects only the length of the power vector p. It means that we use it only to normalize the powers. In this paper α is selected such that the squared length of p equals the number of nodes in the network. Then, $p_i(\alpha, \beta) = 1$ means (approximately) that position i does not have an unusually large or small degree of centrality.

Recall from Principle 3 that c denotes the upperbound of relative degrees of individuals in the network. We use λ to denote the dominating eigenvalue of the adjacency matrix of the weighted influence graph. The parameter β can be any value on the interval $[-\frac{1}{\lambda}, \frac{1}{\lambda}]$ where λ is the largest eigenvalue of the weighted influence graph. When $\beta = 0$, Bonacich power coincides with degree centrality. When $\beta > 0$, a node become more powerful as its neighbors become more powerful. In contrast, when β is negative, nodes become more powerful as their neighbors become less powerful (this is useful to capture negative exchange networks). Intuitively, we use the parameter $0 < \beta < 1$ to characterize the *loss of control* of the managers [9]. The more connections an individual has, the less their power depends on each of their neighbors' powers. Capacity indirectly indicates how much time a worker spends with each of their subordinates, friends or collaborators. Therefore, we assume that β is inversely proportional to the capacity minus one (one is for the relation with the manager):

$$\beta < \begin{cases} \frac{1}{\lambda} \text{ if } \lambda > 1, \\ \frac{1}{c-1} \text{ otherwise} \end{cases} \tag{3.3}$$

3.4 Main Features of CORPNET

CORPNET is a software application for interactive simulation, analysis and visualization based on the network model above. It is developed using the Scala programming language, and runs on the Java Virtual Machine. The program allows user customization over a range of key parameters, whose meaning is as described abvoe:

- β: A scale factor for β, in the range (0, 1)
- Δ: The total available capacity of each node.
- w: The capacity used by each social edge.
- k: The weight of backflow and social edges.
- ρ: The weight of edges between top managers
- h: The maximum height of the reporting hierarchy.

The user interface consists of three parts; a menu bar, a visualization area, and log output area. The internal architecture is multithreaded; all user interface actions trigger an asynchronous message passed to a finite state machine, which in turn executes computations on separate threads. The numerical calculations on the network graph adjacency matrices are handled by `Breeze`, a numerical processing library for Scala, which makes use of native `BLAS`, `LAPACK` and `ARPACK` libraries, thus achieving maximum performance.

The network graph can be visualized as either a *tree layout* or a *force directed layout*. The tree layout is hierarchical with respect to the directed edges only. The color of a node can represent either its power, or its membership within a detected community. See Figs. 3.2 and 3.3 for examples of both layouts.

CORPNET computes Bonacich power as defined above and visualize this information in multiple ways. For both layout styles, a darker shade of blue indicates a higher relative power within the network. The tree layout also displays the power of a node as numeric text. The force layout draws nodes with varying sizes, such

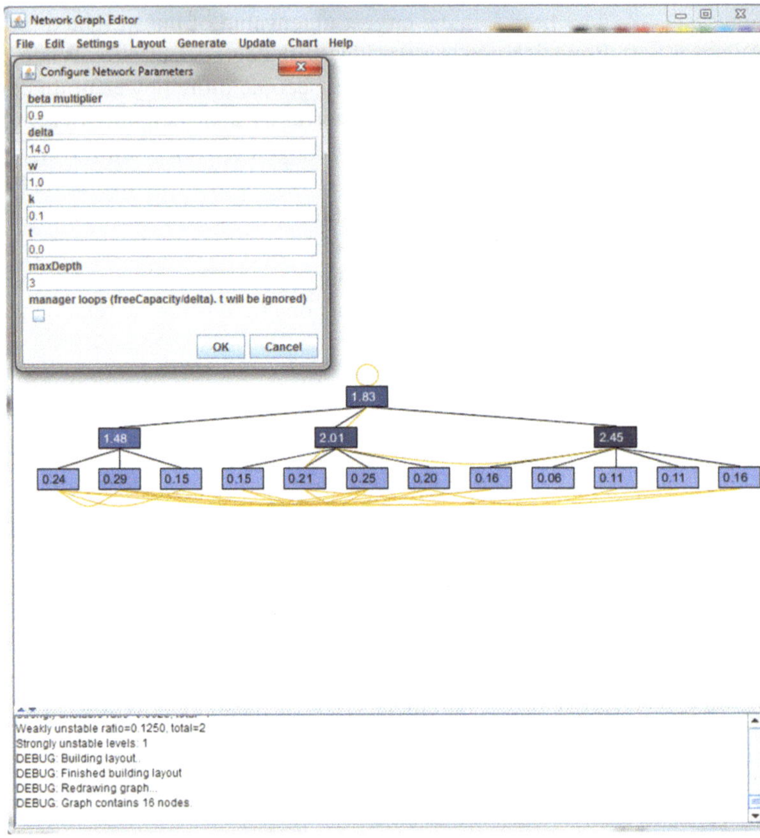

Fig. 3.2 CORPNET user interface: a tree layout example with parameter setting panel

that more powerful nodes are relatively larger. By default, the power values are immediately recalculated for all nodes after the user modifies the network by either adding/removing nodes and edges, or adjusting model parameters. This automatic recalculation can be disabled, which is convenient when making a batch of changes to a large network (it can take a few seconds to recalculate powers with more than a thousand nodes). CORPNET is able to perform ONA from the following perspectives:

Stability Analysis.

We posit that the individual powers influence significantly organizational stability. Intuitively, power of individuals who are closer to the top managers should be greater than the power of individuals further down the hierarchy. This led us to the idea of the stable and unstable networks: an employee in some hierarchy layer l becomes unsatisfied if there is another employee whose power and the hierarchy level are greater. A node is considered *strongly (autocratically) unstable* if it has lower power

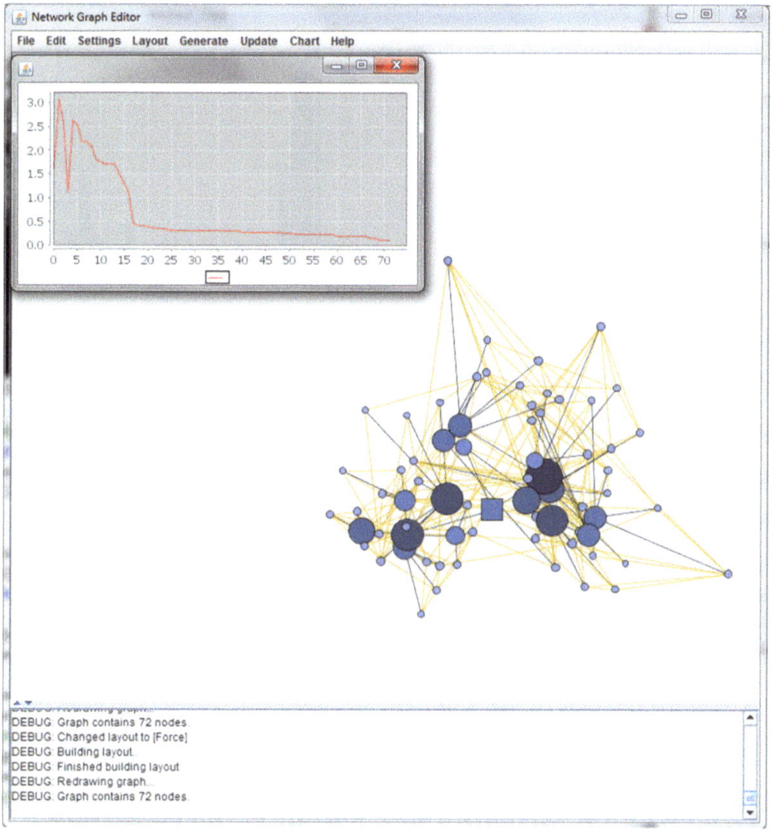

Fig. 3.3 CORPNET user interface: a force directed layout with a distribution of power

than at least one node on a lower level. A node is considered *weakly unstable* if it has lower power than at least one of its descendants. A level is considered strongly unstable if it contains a strongly unstable node.

The program automatically analyzes the stability of the network after calculating the Bonacich power. The analysis result is displayed in the log output area. It shows whether the network is strongly stable, weakly stable. If the network contains any unstable nodes, the report will show the degree of each type of instability, as the ratio of the total number of unstable nodes to the total number of nodes.

Community Detection.

CORPNET incorporates a module which computes communities in the organizations based on formal and informal ties using Newman's spectral algorithm [8]. This is visualized by node colors which represent communities detected.

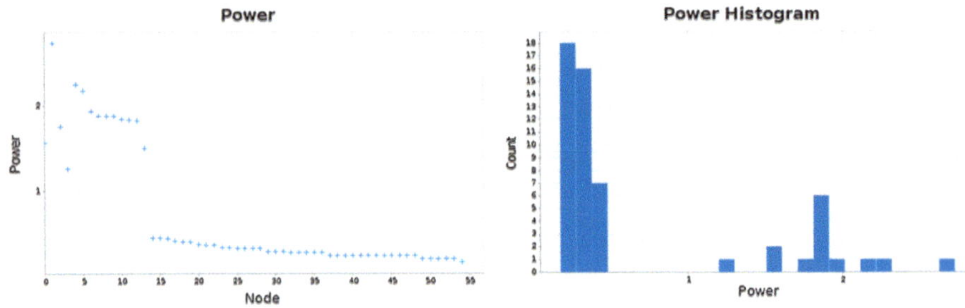

Fig. 3.4 Plots of the distribution of powers: descending power grouped by level (*left*) and Power histogram (*right*)

Power Distribution Plots.

CORPNET offers two types of plots for providing statistical information regarding individual influences (in the menu *Visualize* → *Plot*):

- *Descending power grouped by level*: A scatter plot of individual node powers. The nodes are arranged from left to right in descending order of power, grouped by level such that the nodes on higher levels are to the left of nodes on lower levels.
- *Power histogram*: A histogram of node powers with a configurable number of bins.

See Fig. 3.4 for examples of these plots.

3.5 Organization Structure Simulation Using CORPNET

To facilitate experiments on organizational networks, CORPNET incorporates a network simulator which is able to generate synthetic benchmark organizational networks.

Random Tree Generator.

This module generates a tree with a random number of children per node. The number of children is sampled from a normal distribution with a specified mean and standard deviation. The mean of the normal distribution can be incremented for each level. For example, with *initial mean* 3.0, and *mean increment* 0.5, the first level will have mean 3.0, the second level will have mean 3.5, and so on. If the standard deviation and mean increment are both set to zero, then the result is a perfect tree, with exactly the *initial mean* number of children for all nodes.

Social Network Generator.

This module generates a benchmark social network of undirected edges over the existing hierarchical network. Here, typical benchmark graphs in social network analysis is not appropriate: Homophily is a fundamental social network principle which states that people tend to be linked based on their inherent similarities: social ties are more likely to exist between people with the same background, age, interests. Indeed, numerous management theories argue from a homophily point of view and assert that employees are more likely to establish social connections within a certain "circle", such as departments, offices. Moreover, individuals tend to establish personal ties with those that are at the same or similar levels in the hierarchy [13, 18]. No social network model so far has been defined taking into account this multiplex view of an organization (Existing benchmark graphs such as planted ℓ-partition, relaxed caveman graphs and the LFR graphs [8] are not based on a multiplex model of formal and informal ties). Hence we need to provide our own benchmark graph.

We adopt a distributed approach. We first use the random tree generator describe above to obtain a random tree (i.e., the formal ties of the management network). Then we randomly generate informal ties in the network in the following way: For each node v, iterate over all other nodes u:

1. Compute the *distance* d from u to v in the formal hierarchy, i.e., the number of steps between v and u in the tree (treating all edges are undirected). For example, if u is the manager of v, then their distance is 1, if u are v are in the same level sharing the same manager, their distance is 2.

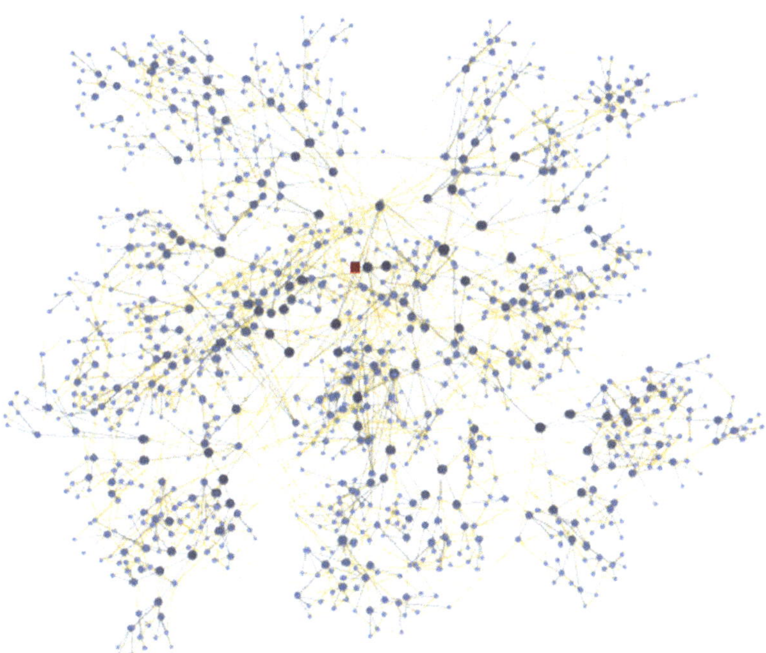

Fig. 3.5 Synthetic benchmark network for $d = 3$ and 7 levels generated by CORPNET. *Blue* and *yellow lines* are formal and informal ties, resp

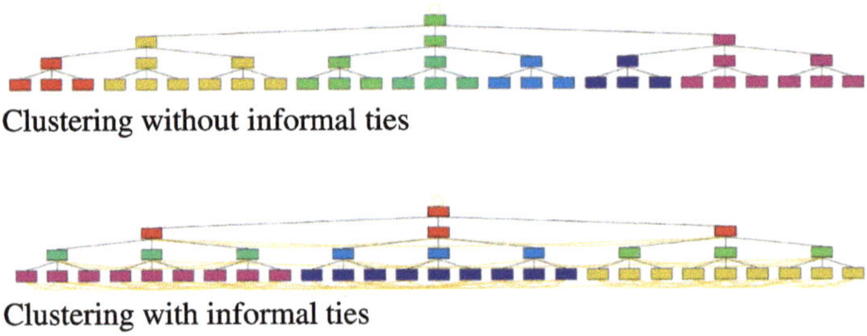

Fig. 3.6 Influence of informal ties on community formations in organizations

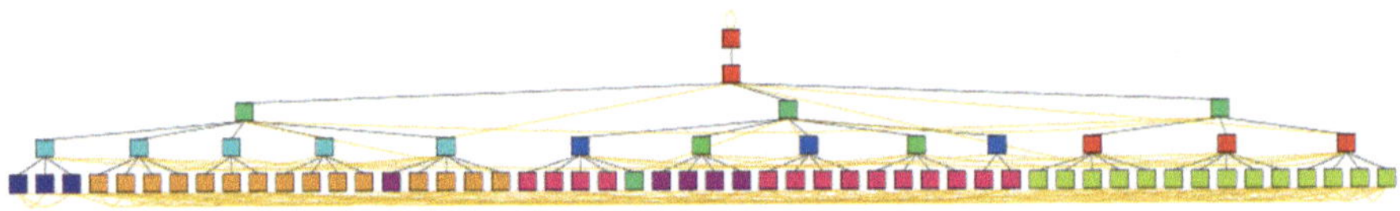

Fig. 3.7 Random tree and random social network with default settings: same colors represent the same clusters

2. Generate an undirected informal tie between u and v using a power law distribution on distance d so that a smaller d means a higher probability of an informal tie between u, v, and a larger d means a lower probability.

The result is a random network that not only captures the main characteristics of social networks (such as community structure), but also entails the reporting hierarchy of the network; see Fig. 3.5 for a generated network visualized using a force-directed method. The community structure clearly resembles departments and reflect hierarchical levels in an organization.

In Fig. 3.6, we first consider a hierarchy with no social network on it. In this case, we can see that clusters reflect departments in some sense, but not the levels of hierarchy. However, when we enrich this hierarchy with a generated social network, we get a quite different picture as in the second picture in Fig. 3.6. The modularity detection algorithms applied to randomly generated trees and social networks as explained above demonstrate that two our propositions hold (see Fig. 3.7):

1. Clusters typically reflect department: people in the same department tend to form a cluster.
2. Clusters typically reveal levels: managers in the same level tend to form a cluster.

3.6 Analyzing Leadership Styles Using CORPNET

A important question in management studies looks at how performance relies on leadership styles in a company [14]. We suggest that CORPNET allows a more rigorous analysis of the connections between personal ties and leadership styles.

Management styles in a company are classified by the level of control exercised by the top managers. For example, managers in an autocratic company make decisions unilaterally with no initiatives from the bottom while in a democratic company, decisions are made by majority rather than by the top managers. The following are main leadership styles adopted by companies and widely studied [17]. Our multiplex network model allows us to capture them in a formal way:

Autocratic Style.

Here the decisions are made from the top management unilaterally. There are few or no initiatives from the bottom of the hierarchy. The number of connections of any individual employee is relatively low because maintaining a reporting relation requires a large amount of resources. Hence we say that an organizational network (with a fixed weighted influence graph) is *autocratic* if $\rho > 1$, k is very small (i.e. within the range $[0, 0, 1]$), and the resources required to maintain informal ties is small ($w \in (0, 0.5)$). The benefit of this style is high stability, while the negative effect is the lack of motivation of employees.

Democratic Style.

Here the decisions are made by majority of the employees. There are many initiatives from the bottom of the hierarchy and collaboration requires as much resources to maintain as the reporting relations. Thus we say that an organizational network is *democratic* if $\rho = 0$, $k \in [0.5, 1)$ and w is close to 1. The benefit of this style lies in job satisfaction and quality, however it does mean a higher level of instability and a inefficiency.

Paternalistic (Consultative) Style.

This leadership style sits somewhere between autocratic and democratic. While the decisions are made mainly from the top managers, they take into account the best interests of the employees. Influence is mainly one-directional (downwards), but feedback also are encouraged. Hence we define an organizational network to be *paternalistic* if $\rho > 0$, $k \in (0.1, 0.5)$, $w \in [0.5, 1)$.

Chaotic Style.

This is a more recent style of management, which gives employees total control over decision making. Here informal ties has become the dominant personal links and thus require a larger weight in their influence and a large amount of resources to maintain. We define an organizational network to be *chaotic* if $\rho = 0$, $k = 1$ and $w \geq 1$. One would expect that any chaotic organizational structure to be unstable.

Figure 3.8 shows three typical networks with different management styles: democratic, paternalistic, and democratic. One can clearly identifies that the distribution of power in such networks is quite different, and reflect the analysis above. We run several experiments by generating different organizational networks with social networks, and get the following results in regards to stability of the networks:

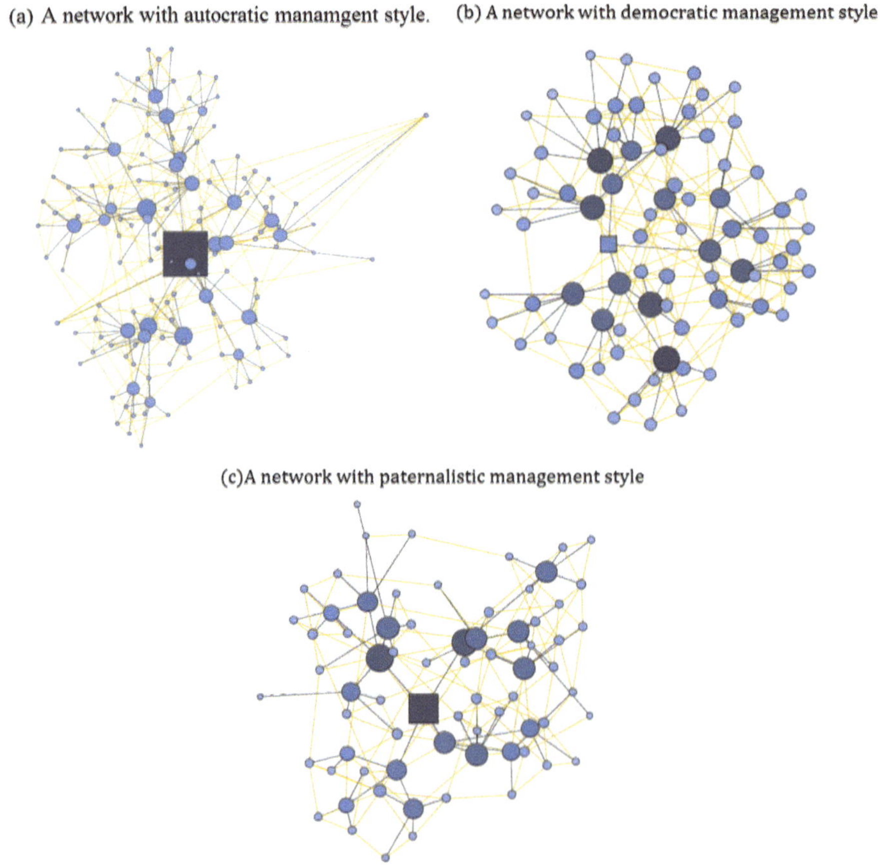

Fig. 3.8 Distribution of power in networks with different management styles

3.7 Conclusion and Future Work

CORPNET is a stand-alone software for visualizing and analyzing the formal and informal ties within an organizational structure. The current prototype of CORPNET incorporates several novel ideas in ONA that include a multiplex network model, and a new benchmark model used to generate informal ties in a corporation, as presented and discussed in detail in [12]. We show how the tool can be use to compute a range of organizational information such as levels of influence, stability and leadership style. Our overall goal is to develop an intelligent decision support system for corporations based on organizational networks which provides useful managerial and strategic advices.

Future work here include enriching the functionalities of the software to more aspects of organization networks. While the current model allows one to discover individual influence, it would be useful to develop a measure of individual effectiveness and performance and investigate how it affects the performance of the overall company. On the other hand, we would also like to investigate detecting "hidden" hierarchies that are not revealed by the formal ties.

References

1. Barney, J., Griffin, R.: The Management of Organizations: Strategy, Structure, Behavior. Houghton Mifflin, Boston (1992)
2. Bonacich, P.: Power and centrality: a family of measures. Am. J. Soc. **92**, 1170–1182 (1987)
3. Calvo, G., Wellisz, S.: Supervision, the firm. J. Political Econ. **86**, 943–952 (1978)
4. Cross, R., Borgatti, S., Parker, A.: Making invisible work visible: using social network analysis to support strategic collaboration. California Manage. Rev. **44**(2), 25–46 (2002)
5. Cross, R., Parker, A.: The Hidden Power of Social Networks. Harvard University Press (2004)
6. Easley, D., Kleinberg, J.: Networks, Crowds, and Markets: Reasoning About a Highly Connected World. Cambridge University Press (2010)
7. Ehrlich, K., Inga, C.: Inside Social Network Analysis. Collaborative User Experience Technical Report. IBM Corporation (2005)
8. Fortunato, S.: Community detection in graphs. tCoRR abs/0906.0612 (2010)
9. Geanakoplos, J., Milgrom, P.: A theory of hierarchies based on limited managerial attention. J. Jpn. Int. Econ. **5**(3), 205–225 (1991)
10. InFlow. http://orgnet.com/inflow3.html
11. Krackhardt, D., Hanson, J.R.: Informal networks: the company behind the chart. Harward Bus. Rev. **71**(4), 104–113 (1993)
12. Liu, J., Moskvina, A.: Hierarchies, Ties and Power in Organisational Networks: Model and Analysis. In: Proceedings of ASONAM 2015, pp. 202-209. ACM (2015)
13. McPherson, M., Smith-Lovin, L., Cook, J.: Birds of a feather: homophily in social networks. Ann. Rev. Soc. **27**, 415–444 (2001)
14. S. sales supervisory style and productivity. Pers. Psychol. **19**, 275–286 (1966)
15. SYNAPP. http://www.seeyournetwork.com/
16. SYNDIO. https://synd.io/
17. Tannenbaum, R., Schmidt, W.: How to choose a leadership pattern. Harvard Bus, Rev (1973)
18. Tyler, J., Wilkinson, D., Huberman, B.: Email as spectroscopy: automated discovery of community structure within organizations. Proceedings of C&T **2003**, 81–96 (2003)
19. Teubner., G.: Flattening the organizational structure: encouraging empowerment or reinforcing control?. In: Research in Organizational Change and Development, pp.147–168. Emerald Group Publishing Limited (2001)
20. Williamson, O.: Hierarchical control and optimal firm size. J. Polit. Econ. **75**, 123–138 (1967)
21. Wulf, J.: The flattened firm—not as advertised, Harvard Business School Working Paper, 12–087, 9 Apr 2012

Chapter 4
Trust Transference on Social Exchanges Among Triads of Agents Based on Dependence Relations and Reputation

Yunevda E.L. Rojas, Diana F. Adamatti and Gracaliz P. Dimuro

Abstract This paper proposes the addition of a third agent to the model proposed by Piaget to Social Exchanges. This agent, called an Intermediary Agent, may outsource a service in favor of another agent. This situation can be used to develop a social exchange model for triads of agents in order to evaluate the process of trust transfer between agents who are not familiar with each other, i.e., they have never interacted directly, just by the intermediary agent. The trust transfer analysis is based on exchange value, reputation and dependence relationship concepts, all of which are analyzed interdependently. This way, it complements the Piaget model, generating a contribution to the study of non-economic aspects of the exchange process. The generated information can be used to guide agents' decisions about partner choice in future exchanges. Additionally, it provides a starting point to understanding the agents' behavior in interactions among more than three agents.

Keywords Social exchanges · Trust transference · Dependence relations · Reputation

4.1 Introduction

On the field of the multi-agent systems, the analysis of human society interactions models and theories shows that there are many ways of interacting, that can happen at different levels. For example, it can be an exchange of information, a negotiation, a discussion, the development of shared visions about an environment, or even the formation or dissolution of organizational structures.

Y.E.L. Rojas (✉) · D.F. Adamatti · G.P. Dimuro
Centro de Ciências Computacionais Rio Grande/RS, Universidade Federal do Rio Grande—FURG, Osorio, Brazil
e-mail: yuniekita@gmail.com

D.F. Adamatti
e-mail: dianaada@gmail.com

G.P. Dimuro
e-mail: gracaliz@gmail.com

Q. Bai et al. (eds.), *Multi-agent and Complex Systems*, Studies in Computational Intelligence 670, DOI 10.1007/978-981-10-2564-8_4

The theory developed by [11] is widely used in this field of studies. It models a social exchange as a service exchange interaction between a pair of agents. After each exchange, an evaluation is made by the involved agents, generating the exchange values concept, divided in investment, satisfaction, credit and debit, which are qualitative values. These values help the agents attitude when they need to choose partners for future exchanges. Many studies refer to these interactions, focusing in different points, like coalitions, task delegation, cooperation and coordination between agents [6, 10]. This collective behavior of the agents is used to solve problems in multi-agent systems [15].

Thus, the analysis of the set of exchange values, the relations of dependency, reputation and trust in the approach to social exchanges, allow to understand the subjective values generated in this process of exchanges. The use of the term subjectives values refers to values of qualitative and non-economic nature (for instance, good, bad, trust, etc.), with which everyone judges their daily exchanges.

The subjective values, in comparison to the classical exchange based on economic values and quantitative utility functions (for instance, the amount of utility to be obtained from an exchange), usually cannot be faithfully represented quantitatively due to the lack of objective conditions for the measurement.

In this paper, the concept defined by Piaget is expanded for social relations composed by exchanges among three agents, by adding an Intermediary Agent, that acts as both requester and supplier of a service. The intermediary Agent is characterized by its impossibility of performing a service, requested as payment for a debit, but compromised to fulfill its duty with the requester, searches for a third agent that is able to perform the requested service, outsourcing the request. To keep the continuity of the social exchange process, it is needed to guarantee the existence of future exchanges and the correct choice of exchange partners. To achieve that goal, besides the exchange values, other concepts are needed, that are the dependence relation between agents and the reputation of each involved agent. Considering that a dependence relation, as stated by [13], happens when an agent wants to achieve certain state, which is its objective, but it does not have the possibilities to achieve. So, it searches for a second agent who has every condition needed to help it achieve its goal. The actions of the second agent will become a resource to make it possible for the first agent to achieve its objective. Rodrigues and Luck [12] study the dependence relations in more details. Reputation is seen as a social tool that has the objective to reduce the uncertainty about unknown individuals before interacting with them, as stated by [1, 2, 4, 8, 9] have done studies in more details. Finally, this work integrates within the multi-agent systems, the basis of social exchange, dependence, reputation and trust [3], by treating them interdependently, never alone. Thus, it allows the development of a social exchange model for triads of agents to evaluate the process of trust transference.

4.2 Proposed Models

In relation to human society, Piaget adopted a relational approach, such that society is defined as a structure in which the relationships between individuals are established by social exchanges [5]. In many of those relationships, it is common to observe the existence of intermediaries who make possible to perform the exchange process. In this reality, an extension of the Piagets Social Exchange theory for situations where triads of agents are related in an exchange is proposed. The analysis of the exchange values generated by the evaluation of the service, besides the concept of dependency relations and reputation, which may enable reliable transfer processes, must also be revised, since new factors and relations may appear when considering 3 interacting agents.

The presented models will contribute to the analysis of the non-economic aspects of social exchange processes, since each of the agents involved in this model meets the basic conditions set out in the design of the exchange value system proposed by Piaget.

4.2.1 Social Exchange with Intermediary Agent (SEIA)

The proposed model, entitled social exchange with intermediary agent ($SEIA$), consider three agents, X, I and Y, with independent characters. Both agent X and agent Y know and they are related with agent I, but they do not know each other. These agents interact in two exchange stages.

4.2.1.1 Stage I

Is composed of two independent processes of exchange. In *Exchange 1*, agent X performs a service for agent I. In *Exchange 2*, agent I performs a service for agent Y. Each interaction generates exchange values, as shown in Fig. 4.1a.

- **Stage I—Exchange 1**:
  1. R_{xi}: *Investment* value of agent X for accomplishing a service for agent I.
  2. S_{ix}: *Satisfaction* value of agent I for the service received from agent X.
  3. T_{ix}: *Debt* value of agent I from the satisfaction about the service received from X.
  4. V_{xi}: *Credit* value that agent X acquired with agent I for the accomplishment of a service.
- **Stage I—Exchange 2**:
  1. R_{iy}: *Investment* value of agent I for accomplishing a service for agent Y.
  2. S_{yi}: *Satisfaction* value of agent Y for the service received from agent I.

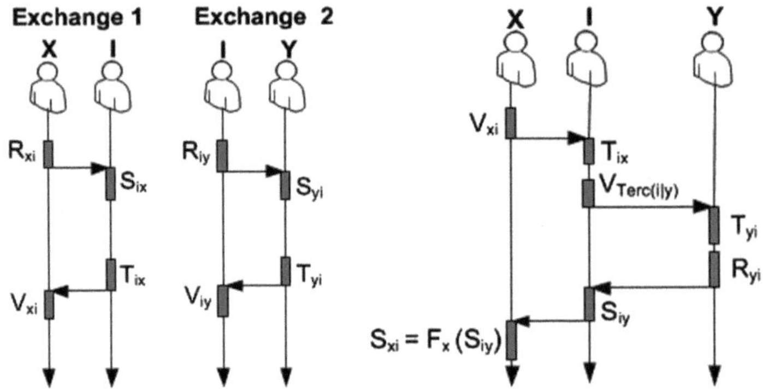

(a) Stage I (b) Stage II

Fig. 4.1 Social exchange with intermediary agent—SEIA

3. T_{yi}: *Debt* value of agent *Y* from the satisfaction about the service received from I.
4. V_{iy}: *Credit* value that agent *I* acquired with agent *Y* for the accomplishment of a service.

Example—Stage I: Exchange 1 and Exchange 2: The TSAI processes are found in different situations everyday. A simple example can take place within a university, where a research project is being conducted between two teachers, and one of them is the advisor of an undergraduate student.

For these examples, in the Exchange 1, the participants are individuals identified as autonomous agents. Thus, at any given time, both the student and the teacher have their own goals and the *investment value* of each of them can be represented for the student as the time and effort in writing the paper. The "service" is represented as the paper. The *credit value* that the student finds himself worthy in delivering the service is represented by the academic advising he wants to receive. The advisor, who receives the service, will generate a *satisfaction value*. A *debt value* will also be generated for the advisor, represented the help that he gives to the student when he needs it.

In Exchange 2, there is the interaction between advisor and teacher in the research paper, the information that the advisor provides to the teacher. The adviser will need some time in the search for information and will take into account the difficulty in selecting information. The teacher will have an expectation and will generate an evaluation of the information received.

4.2.1.2 Stage II

The agent *X* charges a debt from agent I. This charge involves a service concerning its virtual value of credit (V_{xi}), obtained in Stage I. Agent *I* has in its conscience a debt value (T_{ix}) for the received service, but, prevented from the execution of the requested service, outsources it, which consists of the search of a third agent. This interactions generate exchange values, as shown in Fig. 4.1b:

1. V_{xi}: *Credit* value of agent X for accomplishing a service of agent I.
2. T_{ix}: *Debt* value of agent I for the service received from agent X.
3. $V_{terc(i|y)}$: *Outsourcing credit* value from agent I for X in searching for a third agent. I searches for a Y agent, which has a credit value V_{iy} with I, equivalent to T_{ix}.
4. T_{yi}: *Debt* from agent Y to agent I for the satisfaction about the service received.
5. R_{yi}: *Investment* value of agent Y for accomplishing a service for agent I.
6. S_{iy}: *Satisfaction* value of agent I for the service received from agent Y.
7. S_{xi}: *Satisfaction* of agent X for the service received from agent I, generated by the investment of Y for I. The satisfaction value is a function of agent X's evaluation, and is represented by: $S_{xi} = F_x(S_{iy})$.

Example—Stage II—TSAI: In Stage II, the student continues with his research after the conclusion of the paper. He still wants to see if some of his tests are correct. The advisor has to help him guiding the evaluation of the tests required by the student, and the research teacher continues his research activities.

The student has a credit value from the advisor, generated in the previous step. Thus, in one of his daily activities, he asks for help. The advisor recognizes that he has a debt with the student, but he does not have enough knowledge to evaluate the test. So, the supervisor, recognizing that he has a debt to the student, begins a process of outsourcing, in search of another agent with whom he has a credit value and who knows how to solve the problem that the student needs.

4.2.1.3 Formulations of Exchange Values in SEIA

Each obtained value in the SEIA process is quantified based on the definitions and equations described in this section. Calculation of investment and satisfaction values were inspired by the measurement theory LibQual[1] [7], while credit and debit values were supported by [14] investigations.

Definition 1 Investment values R, Eq. 4.1, represent an action or fact of service performance from one agent to other. It is a belief and their values are given by one least the ratio between investment adequacy (A_{inv}) and investment acceptance (Za_{inv}).

$$\mathrm{R} = 1 - \frac{A_{inv}}{Za_{inv}} \tag{4.1}$$

[1] Available at http://www.libqual.org/.

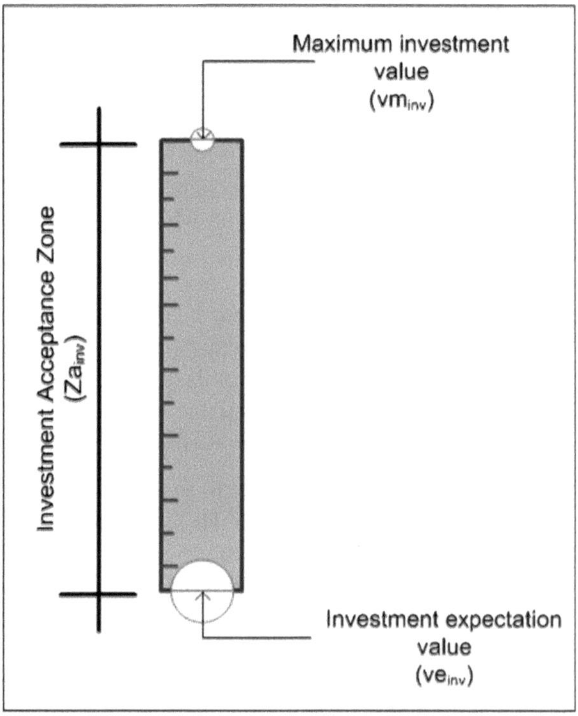

Fig. 4.2 Investment acceptance zone

where:

1. **Investment Acceptance Zone** (Za_{inv}): As shown in Fig. 4.2, it is the distance between the maximum investment value (vm_{inv}) and expected investment value (ve_{inv}). Thus: $Za_{inv} = vm_{inv} - ve_{inv}$.

   (a) **Maximum investment value** (vm_{inv}): It is an individual and differentiated belief of each agent. This value depends on the agent's characteristics and represents everything that the agent can invest in the fulfillment of a service. It can be calculated as the weighted average of all of its values if there are more than one maximum.

   (b) **Investment expectation value** (ve_{inv}): It is also a particular state of each agent and denotes the minimum possible investment that the agent plans to accomplish. As well as the maximum investment value, it can be calculated by the weighted average of expected minimums if there are more than one value.

2. **Investment adequation** (A_{inv}): Portrays the appropriate space for the agent's investment conditions. It represents the distance between the maximum investment value vm_{inv} and the observed investment value vo_{inv}, as follows: $A_{inv} = vm_{inv} - vo_{inv}$ (see Fig. 4.2).

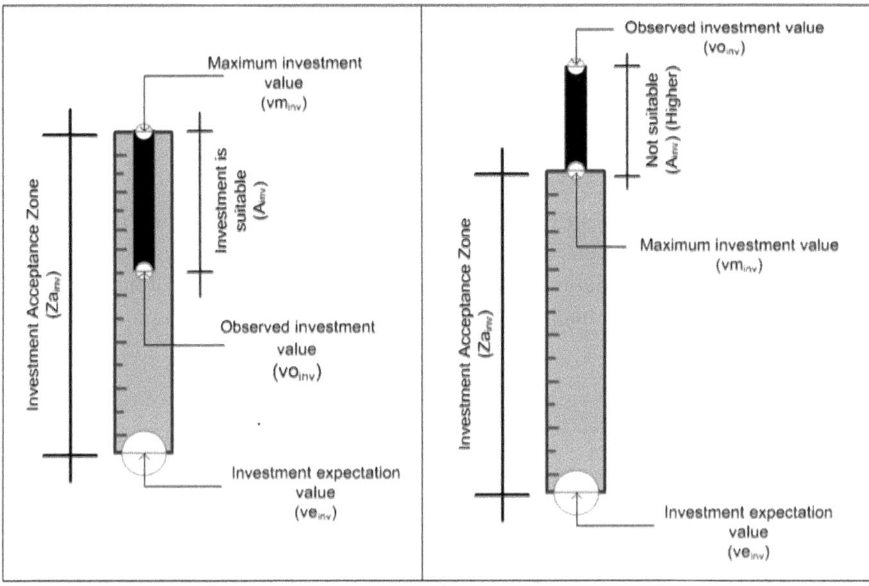

Fig. 4.3 Investment suitable and not suitable

(a) Observed investment value vo_{inv}: The invested value that the agent realizes while carrying out the required actions to achieve the service. The calculation of vo_{inv} is represented by the investment factors weighted average $fi_1, fi_2, fi_3, \ldots, fi_n$ and the weights (pi): $pi_1, pi_2, pi_3, \ldots, pi_n$. Each factor is a completed action and each weight is the importance factor that an agent gives to that factor, obtaining: $vo_{inv} = (pi_1 \times fi_1) + (pi_2 \times fi_2) + \cdots + (pi_n \times fi_n)/(pi_1 + pi_2 + \cdots + pi_n)$. Both factor values and factor weights are linked to particular characteristics of each agent, being different for each one.

When the agent's observed investment level is less than the maximum investment value it can perform, the investment is suitable. If the agent's observed investment level is more than the maximum investment value, the investment is not suitable, as shown in Fig. 4.3. In this case, the investment will not be made, because it overcomes the offer possibilities. Thus, the exchange will not be fulfilled. When the agent's observed investment is less than the expected value, the investment is not significant, but the exchange process must continue.

Definition 2 The virtual credit value, V, resultant from performed services, is the value that the agent finds itself worthy of return from the provision of a service it has performed. Depending on the scenario, the agent can valorize or depreciate its investment. This variation depends on the characteristics of each agent, which for the formulation is represented by an appreciation or depreciation rate of the amount invested, k_{inv}. This rate assists in calculating the virtual value of credit that the agent finds itself worthy, as presented in Eqs. 4.2 and 4.3:

$$Valorization : \mathrm{V} = R + (1 - R)k_{inv} \tag{4.2}$$

$$Depreciation : \mathrm{V} = (1 - k_{inv})R \tag{4.3}$$

Definition 3 The satisfaction value, S, is the resulting pleasure of the performance of what was expected or desired from the received service. It is calculated as the average between the adequation of the service A_{serv} and the zone of acceptance of the service Za_{serv}, as specified in Eq. 4.4.

$$\mathrm{S} = \frac{A_{satisf}}{Za_{satisf}} \tag{4.4}$$

where:

1. **Satisfaction acceptance zone** (Za_{satisf}): As shown in Fig. 4.4, it is the space represented by the distance between the minimum satisfaction value, vm_{satisf}, and the desired satisfaction value, vd_{satisf}. This way: $Za_{satisf} = vd_{satisf} - vm_{satisf}$. Every agent will accept what is above its minimum satisfaction value to its desired satisfaction value, generated after receiving a service.

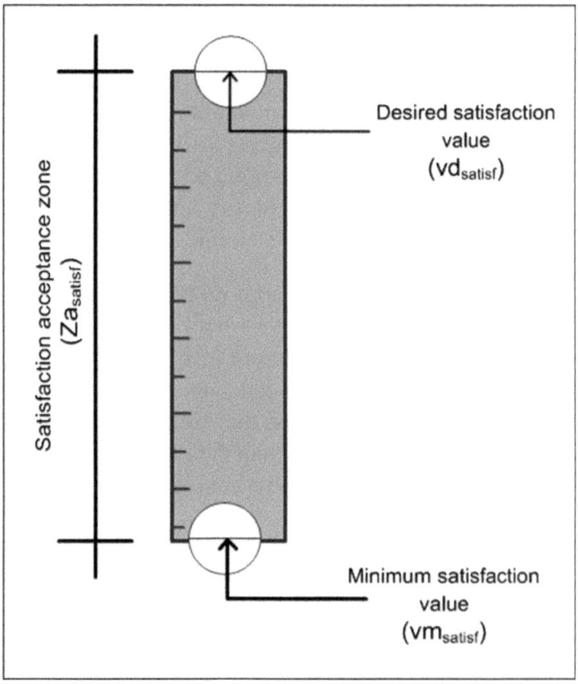

Fig. 4.4 Satisfaction acceptance zone

(a) **Desired satisfaction value** (vd_{satisf}): It is an individual and differentiated belief for each agent. This value depends on qualities and characteristics of the agent. It represents the agent expectative about a service it is about to receive. It can be calculated as the weighted average of all of its maximum values if there are more than one.
(b) **Minimum satisfaction value** (vm_{satisf}): It is a particular condition in each agent and the value that makes the service acceptable. Any value below it causes the service to be considered unacceptable for the agent, disabling the possibilities of the exchange process occurs. It is possible to calculate the weighted average of the minimum values if there are more than one value.

2. **Satisfaction adequation** (A_{satisf}): It means that the satisfaction is appropriate to the prevailing conditions and circumstances. It is calculated as the difference between the observed value and the minimum value, as follows: $A_{satisf} = vo_{satisf} - vm_{satisf}$.

(a) **Observed satisfaction value** (vo_{satisf}): It is the satisfaction that the agent generates when a service is received. It is calculated as the weighted average of factors, $fs_1, fs_2, \ldots, fs_n$, and weights, $ps_1, ps_2, ps_3, \ldots, ps_n$. Each factor has an action to evaluate and weights represent the degree of importance of each factor to the agent, as shown in: $vo_{satisf} = (ps_1 \times fs_1) + (ps_2 \times fs_2) \cdots + (ps_n \times fs_n)/ps_1 + ps_2 + \cdots + ps_n$. Both the value of the factors and the weights are attached to particular characteristics of each agent, differing for each agent.

When the agent satisfaction level is above the minimum value, then the service is adequate. If the opposite occurs, the value of satisfaction obtained is below the minimum value, then the service, according to the agent, is not suitable, as shown in Fig. 4.5. In this situation, the service will no longer be considered a service, and it will not carry out the exchange process between the agents. If the user-perceived value is higher than the desired value, then the service is considered of superior quality.

Definition 4 The virtual value of debt, T, generated as a result from the satisfaction about the services received, is the value that the agent thinks that is its debt. In certain scenarios, the agent can appreciate or depreciate the received value of satisfaction, influencing the final output value. This variation of satisfaction depends on the characteristics of each agent and is represented as a rate, k_{satif}, of appreciation or depreciation of the value of satisfaction, as shown in Eqs. 4.5 and 4.6:

$$Valorization : \mathrm{T} = S + (1 - S)k_{satif} \tag{4.5}$$

$$Depreciation : \mathrm{T} = (1 - k_{satif})S \tag{4.6}$$

Definition 5 The credit outsourcing value, $V_{terc(xiy)}$, is generated when there is a debt whose value can not be paid by the agent itself. In this case, the agent will look for another agent with whom it has a credit value, which must be equivalent to the

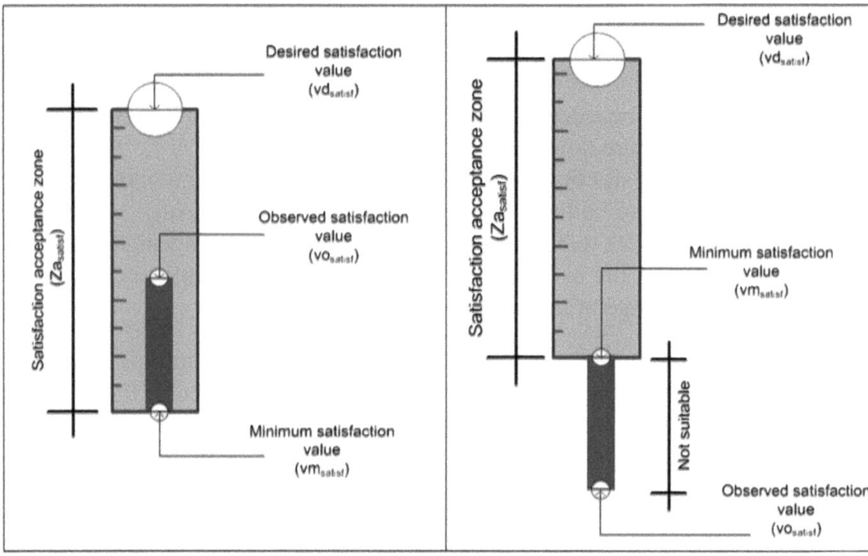

Fig. 4.5 investment is suitable and not suitable

debit value it has with X, to request the performing of the requested service. This behavior is shown in Eq. 4.7:

$$V_{terc(i|y)} = \begin{cases} V_{iy} \ If \ (\exists y)(T_{ix} \cong V_{iy}) \\ 0 \ If \quad (\nexists y) \end{cases} \tag{4.7}$$

4.2.2 Dependence in SEIA

The agents X, I and Y, which are interacting in services exchanges, produce exchange values, (r, s, t, v), for each interaction, called exchanges values belief memory. In addition, each agent has a list of services it can offer. This generates between the agents explicit and implicit dependencies, as presented in Fig. 4.6. The explicit dependencies are created between agents who know each other and they had some type of exchange interaction in the provision of services, like X-I and I-Y. The implicit dependencies are generated between agents who knows each other, X-Y, but they have done some interaction in the exchange of services through an intermediary agent, who interacted with both of them.

In addition, there may be different degrees of dependency between the different agents: **Very weak dependencies**, characterized by an agent who has no credit to claim. It needs an agent to perform a service that it could make, since it can do the job. **Weak dependencies**, characterized by an agent who has a credit to charge and needs to perform a service that it can accomplish, as it has the knowledge to perform

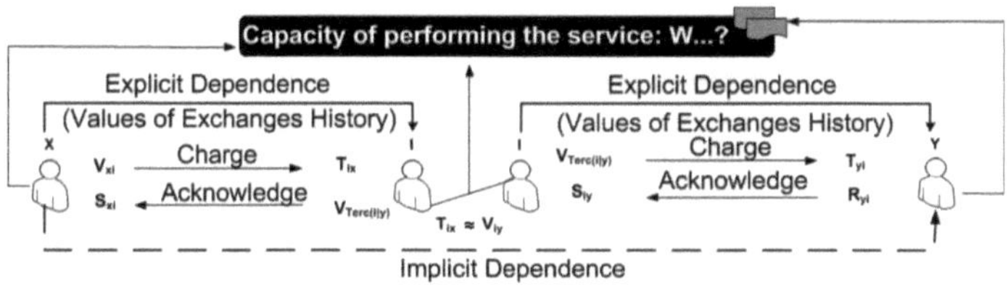

Fig. 4.6 Dependence relations in SEIA

the service. **Strong dependencies** are characterized by an agent who does not have a credit to charge and needs to perform a service that it cannot accomplish by itself, depending of the other agent knowledge. Finally, **very strong dependencies**, are those where the agents have a credit to earn and it needs a service but it does not have the knowledge to perform.

The very weak dependencies and the weak dependencies have the particularity of ceasing to be dependencies, because independent of whether or not there is a credit charge, the agent itself can perform the service, if necessary, to achieve its goal, but decided to ask another agent, based on certain circumstances. Note that the dependence degree varies and that the higher the credit value, the stronger is the dependence and the lower is the power of the agent to perform the service, the stronger is the dependence it has. The degree of dependence is calculated taking into account the dependence in Definition 6.

Definition 6 The degree of dependence represents numerically how much an agent depends on the other. This calculation takes in account the credit value V, and the capacity of performing the service value W. The capacity of performing the service value varies between 0 and 1. Zero (0) represents that the agent is unable to do the service and need another agent for the service to be performed. One (1) represents that agent which has all the conditions to perform the service it needs, so it is up to it to decide whether to outsource the service of not. Values between 0 and 1 indicate different levels of the capacity of performing the service. Equation 4.8 represents the degree of dependence.

$$G_{dep} = V + (1 - V)(1 - W) \tag{4.8}$$

4.2.3 *Reputation in SEIA*

The reputation model in SEIA, as shown in Fig. 4.7, consists of three agents X, I and Y, and it is characterized by the fact that each agent is making direct or indirect evaluation of the services it receives in each interaction of the SEIA. Thus, the reputation of an agent depends on its exchange values history and also on the knowledge of the existing dependence degree.

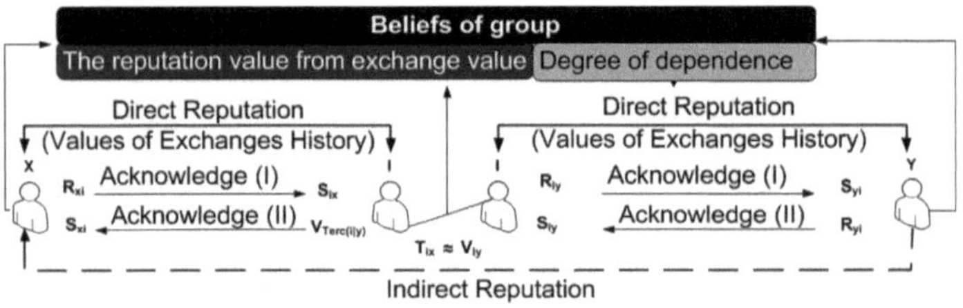

Fig. 4.7 Reputation in SEIA

Construction of Beliefs in SEIA: The construction of beliefs about the reputation value of exchange values, $Vrep$, is a product of the comparison between the observed satisfaction value desired satisfaction value, in the provision and contra-provision of services in the first and second stages of the exchange, respectively, as detailed below:

- I evaluates the service performed by agent X in Stage I—Exchange 1:
  - I generates a satisfaction value about the service received, product of the investment made by agent X.
  - I analyses X by comparison in the trichotomy of the observed satisfaction value in the reception of the service and the desired satisfaction, as follows: $S_{ix} > vd_{satisf}$; $S_{ix} = vd_{satisf}$ and $S_{ix} < vd_{satisf}$, to obtain a reputation value through the exchange values $V_{Rep} \in \{1; 0.5, 0\}$, respectively.
  - I makes a record about the reputation values obtained from the trichotomy about the exchange it made with X.
- Y evaluates the service made by agent I in Stage I—Exchange 2:
  - Y generates a satisfaction value about the service received, product of the investment made by agent I.
  - Y analyses I by comparison in the trichotomy of the observed satisfaction value in the reception of the service and the desired satisfaction, as follows: $S_{yi} > vd_{satisf}$; $S_{yi} = vd_{satisf}$ and $S_{yi} < vd_{satisf}$ to obtain a reputation value through the exchange values $V_{Rep} \in \{1; 0.5, 0\}$ respectively.
  - Y makes a record about the reputation values obtained from the trichotomy about the exchange it made with I.
- I evaluates the service delivered by agent Y in Stage II:
  - I generates a satisfaction value in the reception of the service, product of the investment made by agent Y.
  - I analyses Y by comparison in the trichotomy of the observed satisfaction value in the reception of the service and the desired satisfaction, as follows: $S_{iy} > vd_{satisf}$; $S_{iy} = vd_{satisf}$ and $S_{iy} < vd_{satisf}$, to obtain a reputation value through the exchange values $V_{Rep} \in \{1; 0, 5; 0\}$, respectively.
  - I makes a record about the reputation values obtained from the trichotomy about the exchange it made with Y.

- X evaluates the service performed by agent I in Stage II:
  - X generates a satisfaction value about the service received, product of the investment made by agent I.
  - X analyses I by comparison in the trichotomy of the observed satisfaction value in the reception of the service and the desired satisfaction, as follows: $S_{xi} > vd_{satisf}$, $S_{xi} = vd_{satisf}$ and $S_{xi} < vd_{satisf}$, to obtain a reputation value through the exchange values $V_{Rep} \in \{1; 0, 5; 0\}$, respectively.
  - X makes a record about the reputation values obtained from the trichotomy about the exchange it made with I.

Direct and indirect evaluations are represented as direct reputations and indirect reputations, respectively. So, direct reputation comes from agents who know each other, and indirect reputation comes from agents that do not know each other directly, but exchanged a service indirectly. Figure 4.7 shows how reputation is distributed in a SEIA. Agent X generates a direct reputation (Rep_d) about agent I, and an indirect reputation (Rep_i) about agent Y. Agent I generates direct reputations, (Rep_d), about agents X and Y. Finally, agent Y generates a direct reputation, (Rep_d), about agent I.

Besides the reputation value from exchange values, V_{Rep}, the degree of dependence, G_{dep}, is also part of the calculation of the reputation of an agent, as stated in the Definition 7.

Definition 7 Each agent X, I, Y will have a neutral initial reputation specified for each system. The reputation value, Rep, defined below, is the balance of the evaluation that the agent will get from other agents, as shown in Eqs. 4.9:

$$Rep = \begin{cases} V_{Rep} + (1 - V_{Rep}) \times G_{dep} & If: G_{dep} \geq 0, 5 \wedge V_{Rep} \in \{1; 0, 5\} \\ (1 - G_{dep}) \times V_{Rep} & If: G_{dep} \geq 0, 5 \wedge V_{Rep} = 0 \\ V_{Rep} & If: G_{dep} < 0, 5 \wedge V_{Rep} \in \{1; 0, 5; 0\} \end{cases} \tag{4.9}$$

Finally, the verbal equivalent of the reputation value will belong to the set of positive real numbers ranging from 0 to 1, considering the reputation as bad when it is in range [0; 0, 25), that is:

$$Rep = \begin{cases} Excellent & \Longleftrightarrow Rep = 1 \\ Good & \Longleftrightarrow Rep \in [0, 75; 1) \\ Regular & \Longleftrightarrow Rep \in [0, 5; 0, 75) \\ Bad & \Longleftrightarrow Rep \in [0, 25; 0, 5) \\ Terrible & \Longleftrightarrow Rep \in [0; 0, 25) \end{cases}$$

4.2.4 Trust Transfer in SEIA

The trust transfer process in SEIA, as modeled in Fig. 4.8, consists of three agents X, I and Y. Agent X, who does not know agent Y, will evaluate the possibility to transfer its trust to an agent with whom it had no direct interactions, but has information about it.

Agent X about Agent I:

1. X has a potential goal (V_{xi}) to reach in certain circumstances (T_{ix}). X **believes** that, if it wants to reach that goal and circumstances do not change, then:
   (a) I will be able to make an action R_{ix}.
   (b) I, performing the action R_{ix} ensures to reach the goal V_{xi}, obtaining a satisfaction value S_{xi} desired by X.
   (c) I has the intention of performing R_{ix} because it has a debt with X.

Agent I about Agent Y: Since I cannot perform the service requested by X, it searches for a third agent, Y, who is able to do it. The same that happened between X and I will happen again between I and Y.

1. I has a potential goal $V_{terc(xiy)}$ to reach in certain circumstances (T_{yi}). I **believes** that, if it wants to reach its goal of paying its debt with X and the circumstances do not change, then:
   (a) Y will be able to perform an action R_{yi}.
   (b) Y, performing the action R_{yi}, ensures to reach the I goal: $V_{terc(xiy)}$, and, that way, through I, Xs objective: V_{xi}. This provides the satisfaction values S_{iy}, desired by I, and S_{xi}, desired by X.
   (c) Y has a predisposition to perform R_{yi}, because it has a debt with I.

Agent X about Agent Y:

Based on the trust transfer, agent X will believe in Y after getting the payment of the credit requested from agent I and embodied by Y. It can be said that:

1. If X has a potential goal (V) to achieve in certain circumstances (T), it believes that if it wants to achieve that goal and the circumstances do not change, then:

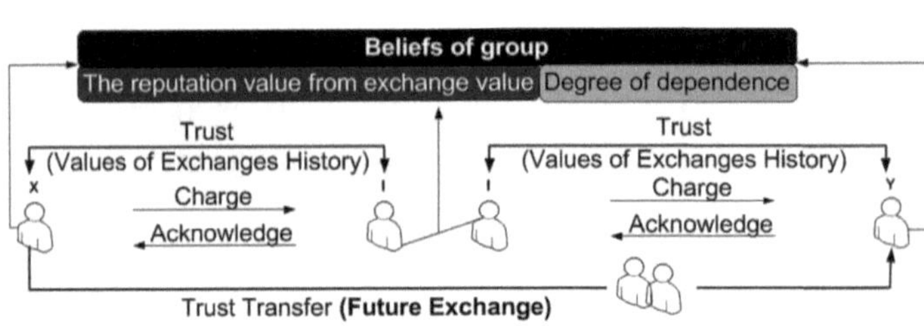

Fig. 4.8 Trust transfer relations

(a) Y will be able to perform an action R.
(b) Y, by performing the action R, ensures to reach X goal, represented by the credit value, V, that it wants to obtain, generating also the desired satisfaction value S.
(c) Y has the intention to make an investment, R, as part of a social exchange process.

This requires developing an analysis of the group's believe. These believes are composed by the reputation of each agent, whereas the reputation value used is calculated taking into account the historical changes of values and relations of dependency, as described in the Definition 7 about building reputation believes in SEIA.

Definition 8 In a trust transference process, an X agent can transfer his trust to a Y agent that he never met if, and only if, there is an I agent, known by X and Y. Besides, for the X agent, the final reputation of I, which provides information about Ys reputation, must be **greater of equal** to R^x_{acep}, which is the acceptable reputation value set by X. After the fulfillment of this condition, X evaluates the final reputation of Y set by I, verifying again if it is greater or equal to X acceptable reputation, as shown in Eq. (4.10).

$$\begin{aligned} If : [(\frac{\sum_{i=1}^{n} Rep_{I_i} \vee Rep_{I_{i+1}} \vee, \ldots, \vee Rep_{I_n}}{n} \geq R^x_{acep}) \\ \bigwedge(\frac{\sum_{i=1}^{n} Rep^y_{I_i} \vee Rep^y_{i+1} \vee, \ldots, \vee Rep^y_{I_n}}{n} \geq R^x_{acep})] \\ \Rightarrow V^{x \rightarrow y}_{transf} \end{aligned} \tag{4.10}$$

where:

- $V^{x \rightarrow y}_{transf}$: Trust transfer from X, with respect to Y.
- n: Total number of agents that provide information.
- $Rep_{I_1} \vee Rep_{I_2} \vee, \ldots, \vee Rep_{I_n}$: Final reputation values of the agents I_1 and I_2 and ... and I_n, which know agent Y and interacted with agent X.
- $Rep^y_{I_1} \vee Rep^y_{I_2} \vee, \ldots, \vee Rep^y_{I_n}$: Final reputation values provided by each of the agents I_1 or I_2 or ... or I_n, about agent Y.
- R^x_{acep}: Acceptable reputation value set by agent X.

It is important to notice that the acceptable reputation value set by x is a function of its intrinsic characteristics.

For an interaction in triads of agents: X, I, Y:

1. X wants to interact with unknown agent Y to reach its target in certain circumstances:
    - X verifies the reputation value provided by I about Y: Rep^y_i.
2. X **transfers its trust** to Y, by **trusting** in I to reach its target when:
    - X, to rely on I, checks and compares if the reputation it has about Rep^i_x is greater than or equal to the reputation value that it considers acceptable, as follows: $Rep^i_x \geq R^x_{acep}$.

3. X, **do not transfer its trust** to Y, **not trusting** in I, when:
   - X checks and compares if the reputation it has about I, Rep_x^i is under the reputation value that it considers acceptable, $Acep$: $Rep_x^i < R_{acep}^x$.

Briefly, X will trust that Y would help it achieve its goal only if it is necessary and the existing information about Y is favorable to its needs. Thus, in a future exchange, X can transfer its trust and relate directly with Y.

4.3 Final Considerations

In the real world, it is very common to find intermediaries in exchange processes. Therefore, it is important to consider the extension of Piaget's theory about Social Exchanges to three agents. The presented models are a beginning for the analysis of non-economic aspects of social exchange processes, and to understand the behavior of agents in interactions with more than three agents, taking into account the concepts of dependence, reputation and trust, interdependently. The integration of the concepts of exchange values, dependence, reputation and trust in the social exchange process in triads of agents enabled to model an internal structure for each agent that interacts in these service exchanges. The information of the exchange values generated by the evaluation of the services will influence the reputation and the existing dependency relationships. This will guide the decision of the agents on the choice of partners with whom they never had direct interaction. Thus, it becomes the focus for the analysis of trust transfer process between agents. The proposed models can be submitted to simulations and evaluations through case studies, to validate them quantitatively.

References

1. Bromley, D.B., Wiley, J., and Sons: Reputation, image, and impression management. Wiley, New York (1993)
2. Castelfranchi, C., Falcone, R.: Principles of trust for mas: cognitive anatomy, social importance and quantification. In: International Conference of Multi-agent Systems (ICMAS), pp. 72–79 (1998)
3. Castelfranchi, C., Falcone, R.: Trust Theory: A Socio-Cognitive and Computational Model, 1st edn. Wiley Publishing (2010)
4. Castelfranchi, C., Falcone, R., Firozabadi, B., Tan, Y.: Special issue on trust, deception and fraud in agent societies. Appl. Artif. Intell. J. 763–768 (2000)
5. Dimuro, G.P., Costa, A.C.R., Palazzo, L.: Systems of exchange values as tools for multi-agent organizations. J. Braz. Comput. Soc. **11**, 27–40 (2005)
6. Farias, G.P., Dimuro, G., Dimuro, G., Jerez, E.D.M.: Exchanges of services based on Piaget's theory of social exchanges using a BDI-fuzzy agent model. In: 11th CBIC, pp. 653–658. IEEE (2013)
7. Herrera-Viedma, E., López-Gijón, J., Avila, B.: Satisfacción de usuarios y calidad: El modelo libqual y su aplicación en bibliotecas españolas e iberoamericanas, pp. 21–23 (2008)

8. Herzig, A., Lorini, E., Hübner, J.F., Vercouter, L.: A logic of trust and reputation. Logic J. IGPL Normative Multiagent Syst. **18**(1), 214–244 (2010)
9. Marsh, S.: Formalising Trust as a Computational Concept. Department of Computing Science—University of Stirling, April, Thesis (1994)
10. Pereira, D., Gonçalves, L., Dimuro, G.P., Costa, A.R.C.: Towards the self-regulation of personality-based social exchange processes in multiagent systems. LNAI, pp. 113–123. Springer (2008)
11. Piaget, J.: Sociological Studies. Routlege, London (1995)
12. Rodrigues, M.R., Luck, M.: Analysing partner selection through exchange values. In: MABS, volume 3891 of LNCS, pp. 24–40. Springer (2005)
13. Sichman, J.S., Demazeau, Y.: A first attempt to use dependence situations as a decision criterion for choosing partners in multi-agent systems. In: Workshop on Decision Theory for DAI Aplicartions (1994)
14. Von Laer, A.: Autorregulação de processos de trocas sociais em sma: um modelo de sociedade de agentes bdi evolucionários e culturais no contexto do jacamo. Universidade Federal do Rio Grande, Master (2014)
15. Woolridge, M.: Introduction to Multiagent Systems. Wiley, New York (2002)

Chapter 5
Exploiting Vagueness for Multi-agent Consensus

Michael Crosscombe and Jonathan Lawry

Abstract A framework for consensus modelling is introduced using Kleene's three valued logic as a means to express vagueness in agents' beliefs. Explicitly borderline cases are inherent to propositions involving vague concepts where sentences of a propositional language may be *absolutely true*, *absolutely false* or *borderline*. By exploiting these intermediate truth values, we can allow agents to adopt a more vague interpretation of underlying concepts in order to weaken their beliefs and reduce the levels of inconsistency, so as to achieve consensus. We consider a consensus combination operation which results in agents adopting the borderline truth value as a shared viewpoint if they are in direct conflict. Simulation experiments are presented which show that applying this operator to agents chosen at random (subject to a consistency threshold) from a population, with initially diverse opinions, results in convergence to a smaller set of more precise shared beliefs. Furthermore, if the choice of agents for combination is dependent on the payoff of their beliefs, this acting as a proxy for performance or usefulness, then the system converges to beliefs which, on average, have higher payoff.

Keywords Agent-based modelling · Many-valued logics · Belief aggregation · Consensus

5.1 Introduction

Reaching a consensus by agreeing a shared viewpoint or position is a fundamental part of many multi-agent decision making and negotiation scenarios. In this paper we argue that by exploiting vagueness in the form of explicitly *borderline* cases we can define an operator for belief combination which not only allows a population of agents to reach consensus but also results in them adopting, on average, a more

M. Crosscombe (✉) · J. Lawry
Department of Engineering Mathematics, University of Bristol, Bristol, BS8 1UB, UK
e-mail: m.crosscombe@bristol.ac.uk

J. Lawry
e-mail: j.lawry@bristol.ac.uk

Q. Bai et al. (eds.), *Multi-agent and Complex Systems*, Studies in Computational Intelligence 670, DOI 10.1007/978-981-10-2564-8_5

useful set of beliefs. The basic intuition underlying this operator is that conflicting agents can agree to allocate borderline truth values to propositions about which they hold inconsistent beliefs. For example, two individuals, one of which believes that 'Cameron is an effective prime minister' whilst the other believes that 'Cameron is ineffective', may agree, in some circumstances, to adopt the shared view that 'Cameron is borderline effective/ineffective'.

Of course, beliefs about the world do not exist in isolation but inform and influence our decisions and actions. From this perspective, some sets of beliefs are more positive or useful than others, resulting in better long term performance, perhaps by making the individuals concerned richer, happier or just better able to survive. More generally, in a multi-agent context, different beliefs result in different actions, collecting different payoffs. In this paper we present simulation studies which show that implementing our proposed operator across a population of agents, initially holding diverse beliefs, results in convergence to a smaller subset of more precise shared opinions. Furthermore, under the assumption that better performing agents, i.e. those with higher payoff, are more likely to interact to reach consensus, we show that the range of beliefs obtained at steady state are on average better, i.e. have higher payoff, than the agents' initial beliefs. The formalism adopted here is that of Kleene's three valued logic and the operator investigated has been proposed for single propositions in [2] and extended to multi-propositional languages in [5].

An outline of the paper is as follows: Sect. 5.2 gives a brief overview of consensus modelling. Section 5.3 introduces Kleene logic and the three valued consensus combination operator. Section 5.4 describes simulation experiments in which agents are selected at random to form a consensus provided that they are sufficiently consistent with one another. In Sect. 5.5 we introduce a payoff function for beliefs, so that the payoff of a particular set of beliefs acts as a proxy for the performance of an agent holding those beliefs. We then adapt the experiments described in Sect. 5.4 so that the probability of an agent being selected for consensus is proportional to their payoff. Finally, in Sect. 5.6 we give some discussions and conclusions.

5.2 Background and Related Work

A number of models for consensus have been proposed in the literature which have influenced the development of the framework described in this paper. De Groot [3] introduced a model for reaching a consensus involving a weighted, global updating of beliefs, iterating until an agreement is reached. In DeGroot's model, agents assign a weight distribution to the population before forming a new opinion. By applying their assigned weights to the other agents' beliefs, an agent can control the influence that others have on their own beliefs.

As an alternative to DeGroot's model, the Bounded Confidence (BC) model introduced in [4] provides agents with a confidence measure. An agent quantifies their level of confidence in their own opinions and are then able to limit their interactions to those agents who possess similar beliefs if they are highly confident (small bounds), or extend the range of possible interactions if the agents possess low confidence (large bounds). In this model agents do not a priori assign weights to the beliefs of others, but instead determine such weightings based on similarity and on their own confidence levels. This is similar in essence to the inconsistency threshold that we introduce in Sect. 5.3, but applied on an individual basis.

The Relative Agreement (RA) model [7] then extends the Bounded Confidence model to allow agents to assign weights to the beliefs of others by quantifying the extent of the overlap of their respective confidence bounds. By having agents declare a confidence interval for their beliefs, the model then restricts interactions to those pairs of agents with overlapping intervals. Consequently, agents are only required to assess their own beliefs and are not required to make explicit judgements about those of other agents. Deffuant et al. [7] also moved to a model of pair-wise interactions to better capture social interactions of individuals, the latter being a setting in which group-wide updates to beliefs are unintuitive in that they do not reflect typical social behaviour.

A fundamental difference between our approach and the above models is that we use Kleene's three valued logic to represent beliefs in a propositional logic setting, rather than identify opinions with real values or intervals. Perron et al. [2] have shown that through use of a three-state model for networked consensus of complete graphs, nodes converge to a consensus much faster and with greater accuracy when compared to a restrictive binary model. In the sequel we extend this approach to a more general setting involving larger languages and incorporating a measure of payoff for beliefs.

5.3 A Three Valued Consensus Model

In this section we introduce Kleene's three valued logic [10] as a model of explicitly borderline cases resulting from the inherent vagueness of propositions. We adopt a propositional logic setting as follows: Let $\mathscr{L}$ be a finite language of propositional logic with connectives $\wedge$, $\vee$ and $\neg$, and propositional variables $\mathscr{P} = \{p_1, \ldots, p_n\}$. Also, let $S\mathscr{L}$ denote the sentences of $\mathscr{L}$ generated by recursively applying the connectives to the propositional variables in the usual manner. A Kleene valuation then allocates truth values 0 (false), $\frac{1}{2}$ (borderline) and 1 (true) to the sentences of $\mathscr{L}$ as follows:

Definition 1 Kleene Valuations

A Kleene valuation v on $\mathscr{L}$ is a function $v : S\mathscr{L} \to \{0, \frac{1}{2}, 1\}$ such that $\forall \theta, \varphi \in S\mathscr{L}$ the following hold:

- $v(\neg\theta) = 1 - v(\theta)$
- $v(\theta \wedge \varphi) = \min(v(\theta), v(\varphi))$
- $v(\theta \vee \varphi) = \max(v(\theta), v(\varphi))$

The truth table for Kleene valuations are shown in Table 5.1.

Table 5.1 Kleene truth tables

$\neg$	
1	0
$\frac{1}{2}$	$\frac{1}{2}$
0	1

$\wedge$	1	$\frac{1}{2}$	0
1	1	$\frac{1}{2}$	0
$\frac{1}{2}$	$\frac{1}{2}$	$\frac{1}{2}$	0
0	0	0	0

$\vee$	1	$\frac{1}{2}$	0
1	1	1	1
$\frac{1}{2}$	1	$\frac{1}{2}$	$\frac{1}{2}$
0	1	$\frac{1}{2}$	0

It is sometimes convenient to represent a Kleene valuation v by its associated *orthopair* (P, N) [5], where $P = \{p_i \in \mathscr{P} : v(p_i) = 1\}$ and $N = \{p_i \in \mathscr{P} : v(p_i) = 0\}$. Notice that $P \cap N = \emptyset$ and that $(P \cup N)^c$ corresponds to the set of borderline propositional variables.

Kleene valuations have been proposed as a suitable formalism in which to capture explicitly borderline cases as resulting from inherent flexibility in the definition of vague concepts in natural language [11, 12]. For example, consider the proposition 'Ethel is short'. For the concept short, we might identify a lower height threshold $\underline{h}$ below which any height is classed as being *absolutely short*, and similarly there may be an upper threshold $\overline{h}$ above which any height is *absolutely not short*. If Ethel's height lay between $\underline{h}$ and $\overline{h}$ then this would result in a borderline truth value for the statement 'Ethel is short'.

It is important to note that the middle truth value $\frac{1}{2}$ is not intended to represent epistemic uncertainty, but rather explicitly borderline cases resulting from the inherent vagueness of natural language propositions. Hence, if we say that the statement 'Ethel is short' is borderline true/false we are not saying that the truth or falsity of this proposition is unknown. Instead we are indicating that Ethel's height is a borderline case of the predicate short. In order to emphasise the difference between the epistemic and the borderline interpretation of three valued logic it is helpful to think in terms of conditioning. For instance, if we learn that it is *unknown* whether or not Ethel is short, then this provides us with no new information about her height. In contrast, learning that Ethel is *borderline short* does provide us with new information about Ethel's height, namely that it lies on the borderline between short and not short. A more comprehensive discussion of these issues can be found in [1]. A consequence of using this interpretation of the middle truth value is that in the current paper we only model consensus for sets of propositions which admit borderline cases. In other words, our approach can be used for propositions such as 'Ethel is short' but not, for example, for the proposition 'Ethel is strictly less that 1.4 m tall'.

The following three valued consensus operator was described in detail in [5]:

Definition 2 Consensus Operator
Let v_1 and v_2 be Kleene valuations on $\mathscr{L}$ with associated orthopairs (P_1, N_1) and (P_2, N_2). Then the consensus $v_1 \odot v_2$ is the Kleene valuation with the orthopair

$$((P_1 - N_2) \cup (P_2 - N_1), (N_1 - P_2) \cup (N_2 - P_1))$$

Table 5.2 Truth table for the consensus operator

$\odot$	1	$\frac{1}{2}$	0
1	1	1	$\frac{1}{2}$
$\frac{1}{2}$	1	$\frac{1}{2}$	0
0	$\frac{1}{2}$	0	0

The corresponding truth table for this operator is shown in Table 5.2. From this we can see that the operator preserves the non-borderline truth values 0 or 1 except in the case of a direct conflict i.e. when one agent has truth value 1 and the other 0. In this case both agents adopt the middle truth value $\frac{1}{2}$. Alternatively, from Definition 2 we can think of $\odot$ as an operator which initially weakens both opinions so as to remove direct inconsistencies, before then combining them.

We now introduce two measures that will be used throughout the subsequent simulation experiments.

Definition 3 A Measure of Vagueness
Let v be a Kleene valuation on $\mathscr{L}$ with orthopair (P, N) and n propositional variables. Then we measure the vagueness of v by the proportion of propositional variables which it classifies as being borderline. That is:

$$V(v) = \frac{|(P \cup N)^c|}{n}$$

Definition 4 Inconsistency Measure
Let v_1 and v_2 be Kleene valuations on $\mathscr{L}$ with corresponding orthopairs (P_1, N_1) and (P_2, N_2). Then we define the inconsistency measure of v_1 and v_2 to be the proportion of propositional variables which are in direct conflict between the two valuations i.e. $v_1(p_i) \neq \frac{1}{2}$, $v_2(p_i) \neq \frac{1}{2}$ and $v_1(p_i) = 1 - v_2(p_i)$. That is:

$$I(v_1, v_2) = \frac{|(P_1 \cap N_2)| + |(P_2 \cap N_1)|}{n}$$

Table 5.3 shows the inconsistency truth table of two valuations for a propositional variable, highlighting the cases where two valuations are *inconsistent*, and consistent otherwise. We can see that there is a probability of $\frac{2}{9}$ that two valuations will be inconsistent for each propositional variable in the language. In the sequel we will propose a threshold $\gamma \in [0, 1]$ on inconsistency so that valuations v_1 and v_2 can be combined only if $I(v_1, v_2) \leq \gamma$.

Table 5.3 Inconsistency truth table

I	1	$\frac{1}{2}$	0
1	0	0	1
$\frac{1}{2}$	0	0	0
0	1	0	0

5.4 Simulation Experiments Based on Random Selection of Agents

We introduce simulation experiments in order to investigate the convergence properties of the three valued logic operator when implemented across a multi-agent system. The experimental set up is loosely based on those proposed in [7, 9], although our representation of opinions is quite different with beliefs taking the form of Kleene valuations on $\mathscr{L}$, rather than vectors of bounded real numbers.

We will consider two distinct initialisations of the beliefs of a population of agents. The *random three valued initialisation* allocates the truth values 0, $\frac{1}{2}$ and 1 to each agent and each propositional variable at random i.e. with probability $\frac{1}{3}$ for each truth value. In contrast, the *random Boolean initialisation* only allocates the binary truth values 0 and 1, each with a probability of $\frac{1}{2}$. This latter initialisation will be required in Sect. 5.5 in order to directly compare the proposed three valued combination operator with a similar two valued operator. In this section we will use the random three valued initialisation in order to investigate the extent to which the three valued operator results in convergence to a shared set of opinions across the population of agents.

We set a fixed maximum number of 50, 000 iterations.[1] At each time step a pair of agents are selected at random from the population. An inconsistency threshold value $\gamma \in [0, 1]$ is set, so that for any pair of agents with respective valuations v_1 and v_2, if $I(v_1, v_2) \leq \gamma$ then both agents replace their beliefs with the consensus valuation $v_1 \odot v_2$, while if $I(v_1, v_2) > \gamma$ then no combination is performed and both agents retain their original beliefs. For $\gamma = 1$ we obtain what is equivalent to the totally connected graph model described in [2], in which any pair of agents can combine their beliefs, whilst taking $\gamma = 0$ corresponds to the most conservative scenario in which only absolutely consistent beliefs can be combined. The parameters for the simulation experiments are then as follows:

- Population size: 100
- Language size i.e. $|\mathscr{P}| = n$: 5, 10, 50, 100
- Initial beliefs: Random three valued.
- Inconsistency threshold: $\gamma \in [0, 1]$.

Figures 5.1 and 5.2 show the results for the experiments after 50, 000 iterations. In each case the plots show mean values with error bars representing standard deviation across 100 independent runs of the simulation. Figure 5.1 shows the average vagueness determined by taking the mean value of $V(v)$ (Definition 3) across the population. Note that for a random three valued initialisation of beliefs we expect a mean vagueness value of $\frac{1}{3}$ at the start of the simulation. As the threshold γ increases then the average vagueness decreases to zero, so that for $\gamma \geq 0.3$ we are left with almost entirely crisp (i.e. Boolean) opinions. In general the more conservative the

[1] In preliminary experiments we found that 50, 000 was an upper bound on the number of iterations required for the system to reach steady state across a range of parameter settings.

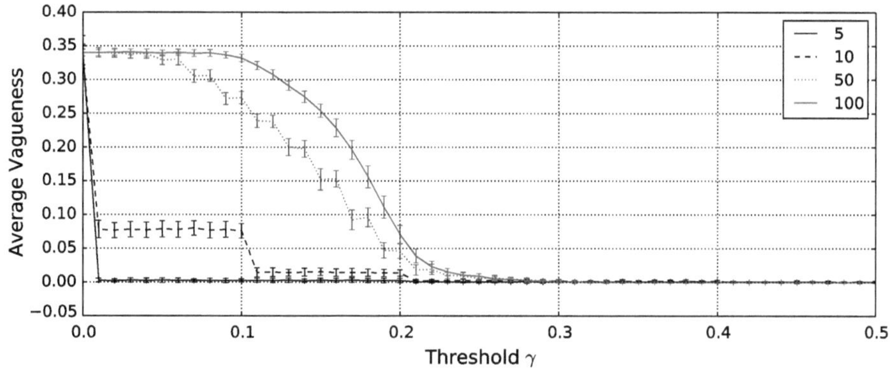

Fig. 5.1 Average vagueness after 50, 000 for varying inconsistency thresholds γ and language sizes

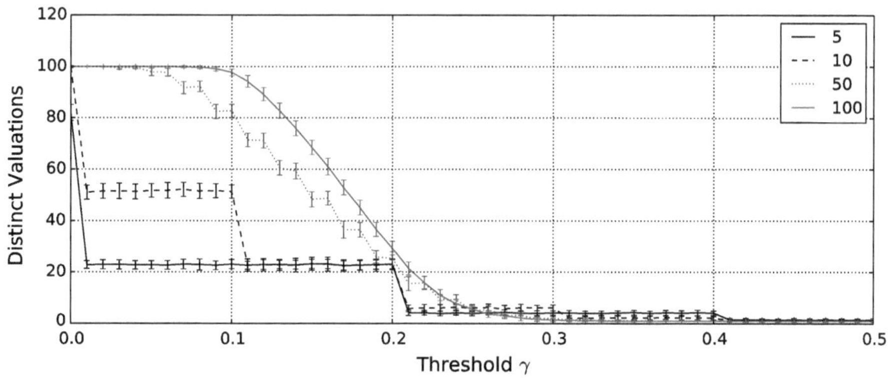

Fig. 5.2 Number of distinct valuations after 50, 000 iterations for varying inconsistency thresholds γ and language sizes

combination rules (i.e. requiring higher levels of consistency) then the more it is that vague beliefs are maintained in the population. Figure 5.2 shows the number of distinct valuations (i.e. different opinions) remaining in the population after 50, 000 iterations. Again this decreases with γ and for $\gamma > 0.4$ agents have on average converged to a single shared belief. This is consistent with the analytical results presented in [2] for the single propositional, $\gamma = 1$ case.

5.5 Simulation Experiments Incorporating a Payoff Model

In this section we extend the simulation framework described in Sect. 5.4 to allow for different payoffs for different beliefs. As outlined in Sect. 5.1, payoff is introduced as a proxy for performance, and is motivated by the intuition that different beliefs result in different actions which then, over time, lead to different levels of performance. Here we adopt an abstract simplification of this process in which each Kleene valuation is allocated a real valued payoff. Then, instead of being selected at random for combination, an agent is picked from the population according to a probability which is proportionate to the payoff value of their beliefs. The idea, then, is that agents with better or more useful opinions will be more successful and furthermore, it will be these successful agents who will be most likely to need to reach a consensus between them.

Here the underlying intuition is that, in real systems it is the most successful agents, with the highest payoff values, who are most likely to find themselves in conflict with one another, and who will most benefit from reaching an agreement. We adopt a simple summative payoff model in which each propositional variable p_i is allocated a value in the range $[-1, 1]$, denoted $f(p_i)$, and the payoff for a valuation with orthopair (P, N) is then calculated as follows:

$$f(P, N) = \sum_{p_i \in P} f(p_i) - \sum_{p_i \in N} f(p_i)$$

Another perspective on this type of payoff function is as follows: For each propositional variable p_i, a truth value of 1 results in a payoff $f(p_i)$ (which can be either positive or negative), a truth value of 0 results in the opposite signed payoff $-f(p_i)$, and a borderline truth value $\frac{1}{2}$ results in a neutral payoff of 0. The payoff value for a Kleene valuation v is then simply taken to be the sum of the payoffs for each propositional variable under the truth values allocated by v.

Based on payoff values we define a probability distribution over the agents in the population according to which the probability that an agent with beliefs (P, N) is selected for possible consensus combination is proportional to $f(P, N) + n$. At each iteration a pair of agents are selected at random according to this distribution. For each such pair the inconsistency measure (Definition 4) is evaluated and either both the valuations are replaced with the consensus valuation, or both are left unchanged, depending on the threshold γ as in Sect. 5.4. The parameters for the simulation experiments are as follows:

- Population size: 100
- Language size: 5
- Initial beliefs: Random Boolean.
- Inconsistency threshold: $\gamma \in [0, 1]$

Notice that here we are initialising the beliefs as random Boolean valuations (see Sect. 5.4).[2] This allows us to make a direct comparison between the performance of the three valued combination operator and a similar two valued operator. For the latter we assume that only binary truth values are available to represent an agent's beliefs. In this context, in order for two agents with conflicting truth values for p_i

[2]As a result of this Boolean initialisation, a language size of 5 now produces a total of 2^5 (32) possible valuations, as opposed to 3^5 (243) possible valuations.

Table 5.4 The truth table for the stochastic Boolean consensus operator

Binary operator	0	1
0	0	$0 : \frac{1}{2}, 1 : \frac{1}{2}$
1	$0 : \frac{1}{2}, 1 : \frac{1}{2}$	1

(i.e. one 0 and the other 1) to reach consensus, we propose that they simply agree to pick one of the truth values at random e.g. by tossing a fair coin. Table 5.4 gives the truth table for the operator in which directly conflicting truth values leads to a stochastic outcome.

The focus on simulations with 5 propositional variables is intended to increase the number of opinions relative to the size of the population, in order to achieve a good distribution of valuations. For example, a language size of 5 allows for 32 possible Boolean valuations. With a population of 100 agents, it is therefore very likely that each opinion will occur at least once. In comparison, a language size of 10 produces 1, 024 possible Boolean valuations which severely decreases the probability of an opinion being present in a population of the same size.

Figures 5.3, 5.4 and 5.5 show the results for simulation experiments with agent selection based on payoff. The results shown are mean values with error bars taken over 100 independent runs of the simulation. Figure 5.3 shows the average population payoff after 50, 000 iterations given as a percentage of the maximal possible payoff value i.e. the payoff for the valuation (P, N) where $P = \{p_i : f(p_i) > 0\}$ and $N = \{p_i : f(p_i) < 0\}$. For both the binary and the three valued operators we show results for simulations in which agents are selected according to payoff (three-valued, Boolean) and at random as in Sect. 5.4 (three-valued random, Boolean random). We see that for all values of γ, the three valued operator with payoff based selection outperforms all of the other approaches. For the former we can also see that average payoff increases with γ. In contrast, for the other approaches, including the payoff operator with payoff based selection, the mean of the average population payoff remains close to 0 after 50, 000 iterations. Figure 5.4 shows the mean number of

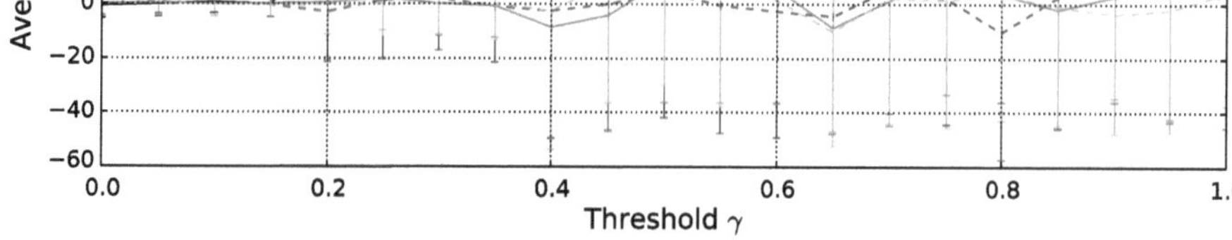

Fig. 5.3 Average payoff after 50, 000 iterations for varying inconsistency thresholds γ, shown as a percentage of the maximal payoff

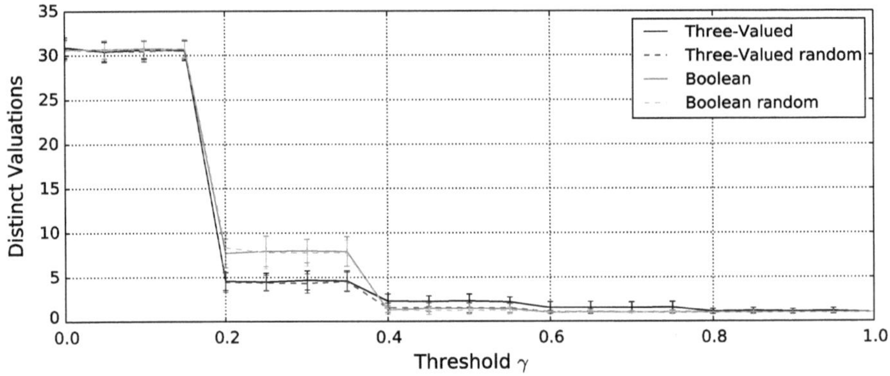

Fig. 5.4 Number of distinct valuations after 50, 000 iterations for varying inconsistency thresholds γ

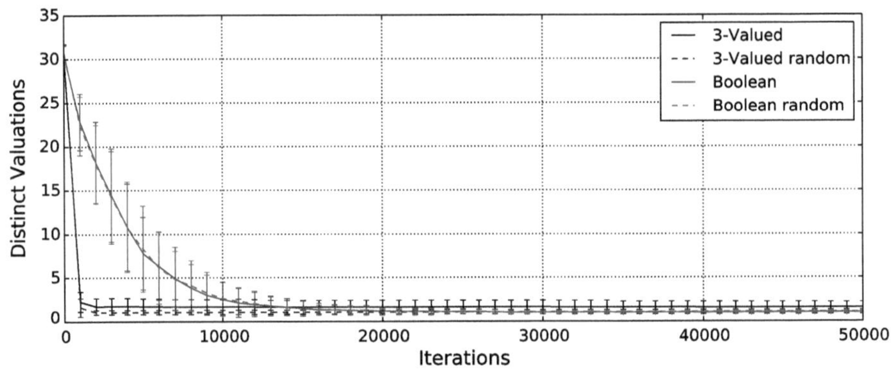

Fig. 5.5 Trajectory showing the number of distinct valuations plotted against iterations for $\gamma = 0.7$

distinct valuations across the population of agents after 50, 000 iterations. All four versions of the operators converge on a small set of shared beliefs for sufficiently large γ. For $\gamma \geq 0.4$ the mean number of distinct valuations is less than 5 while for $\gamma \geq 0.8$ it is 1. Figure 5.5 shows a trajectory of how the number of distinct valuations varies with each iteration when $\gamma = 0.7$. We can see that both the three-valued models converge quickly (in just over 2000) iterations while the Boolean models require considerably longer to converge (over 20, 000 iterations).

5.6 Conclusions

In this paper we have explored the use of Kleene's three valued logic as a framework in which to model multi-agent consensus formation. We have proposed a three valued combination operator, the intuition behind which is that conflicting binary truth values are replaced with a borderline (middle) truth value. A number of simulation experiments have been presented employing this operator. These can be divided into two main categories. For the first type of experiments, agents are selected at random from the population and form a consensus valuation providing that the level of inconsistency of their respective opinions is below a threshold parameter γ. Otherwise they do not form a consensus and instead retain their current opinions. For these experiments we found that there is convergence to a smaller subset of shared opinions across the population. For higher γ values there is convergence on average to a single shared opinion and furthermore this opinion is crisp i.e. it admits no borderlines. For intermediate values of γ the system convergences to a small set of opinions which to some extent remain vague.

In the second type of experiments a payoff function over beliefs is introduced, and agents are selected for possible combination with probability proportional to the payoff value of their current beliefs. Here we compare the three value operator with a similar stochastic Boolean operator. We find that the three valued operator with payoff based agent selection results in convergence to a smaller shared set of beliefs with significantly higher average payoff than that of the initial population. The Boolean operator does not perform well in this context and does not result in a significant increase in average payoff, which instead remains close to 0 after 50, 000 iterations.

The results of the payoff based experiments show how a three valued model for consensus provides a number of improvements over a traditional Boolean model. Firstly, we have shown that the introduction of Kleene valuations to capture the inherent vagueness of propositions does not, in the long run, lead to the mass adoption of borderline truth values as a result of conflict occurring in the population. Instead, we have seen how vagueness is reduced at lower γ values, and at higher γ values the population converges towards completely crisp opinions on average, admitting no borderline cases. In addition to this, we can see that the introduction of a payoff based model drives consensus towards those valuations which result in higher payoff on average. By selecting pairs of agents based on their perceived success, we can achieve an increase to overall payoff in a small number of iterations, compared to no significant increase in payoff for the Boolean model. Therefore, we have shown that the three valued approach incorporating a payoff model can drive convergence across the population towards more successful opinions.

We suggest that the experiments presented in this paper show the potential of using three valued logic in consensus modelling. There is also significant scope to extend the research presented in several new directions. For example, the above studies concern consensus defined at the level of propositional variables. However, in many cases agents will be most concerned to reach agreement about a relevant set of compound statements. For example, they may need to reach agreement about a particular set of conditional statements, or equivalences. Hence, an important question is that of how best to extend our proposed consensus model so as to be applicable to compound logical expressions. Another significant question concerns uncertainty. Suppose that in addition to vagueness agents also quantify their uncertainty about beliefs. Lawry and Dubois [5] propose an extension of the three valued framework in which agents' beliefs are represented by a probability distribution over Kleene valuations. Ongoing research concerns the design of simulation studies in which to evaluate the convergence and payoff based performance of this extended model. Finally, it would be interesting to consider extensions to the operator which allows for consensus between groups rather than just pairs of agents.

Acknowledgments This research is partially funded by an EPSRC PhD studentship as part of a doctoral training partnership (DTP).

References

1. Ciucci, D., Dubois, D., Lawry, J.: Borderline vs unknown: comparing three-valued representation of imperfect information. Int. J. Approx. Reason. **55**(9), 1866–1889 (2014)
2. Perron, E., Vasudevan, D., Vojnovic, M.: Using three states for binary consensus on complete graphs. Technical Report MSR-TR-2008-114. Microsoft Corporation (2008)
3. DeGroot, M.H.: Reaching a consensus. J. Am. Stat. Assoc. **69**(345), 118–12 (1974). JSTOR
4. Krause, U.: A discrete nonlinear and non-autonomous model of consensus formation. communications in difference equations. In: Proceedings of the Fourth International Conference on Difference Equations, pp. 227–237. Gordon and Breach (1998)
5. Lawry, J., Dubois, D.: A bipolar framework for combining beliefs about vague propositions. In: Proceedings of Thirteenth International Conference on the Principles of Knowledge Representation and Reasoning, pp. 530–540. AAAI (2012)
6. Deffuant, G., Neau, D., Amblard, F., Weisbuch, G.: Mixing beliefs among interacting agents. In: Advances in Complex Systems, vol. 3. ACS (2000)
7. Deffuant, G., Amblard, F., Weisbuch, G., Faure, T.: How can extremism prevail? A study based on the relative agreement interaction model. J. Artif. Soc. Soc. Simul. **5**(4) (2002)
8. Hegselmann, R., Krause, U.: Opinion Dynamics and bounded confidence: models, analysis, and simulation. J. Artif. Soc. Soc. Simul. **5**(3) (2002)
9. Meadows, M., Cliff, D.: Reexamining the relative agreement model of opinion dynamics. J. Artif. Soc. Soc. Simul. **15**(4) (2012)
10. Kleene, S.C.: Introduction to Metamathematics, 1st edn. North-Holland (1952)
11. Lawry, J., Tang, Y.: On truth-gaps, bipolar belief and the assertability of vague propositions. Artif. Intell. **191**, 20–41 (2012)
12. Lawry, J., Gonzlez-Rodrguez, I.: A bipolar model of assertability and belief. Int. J. Approx. Reason. **52**(1), 76–91 (2012)

Chapter 6
Selecting Robust Strategies Based on Abstracted Game Models

Oscar Veliz and Christopher Kiekintveld

Abstract Game theory is a tool for modeling multi-agent decision problems and has been used to great success in modeling and simulating problems such as poker, security, and trading agents. However, many real games are extremely large and complex with multiple agent interactions. One approach for solving these games is to use abstraction techniques to shrink the game to a form that can be solved by removing detail and translating a solution back to the original. However, abstraction introduces error into the model. We study ways to analyze games that are robust to errors in the model of the game, including abstracted games. We empirically evaluate several solution methods to examine how robust they are for abstracted games.

Keywords Game theory · Agent based simulation · Agent based modeling · Abstraction

6.1 Introduction

Game theory is widely used for analyzing multi-agent decision problems including auctions [12], security [1, 17], Poker [14], and many others. A game-theoretic analysis begins by specifying a formal model of the decision problem, including the actions players can choose, the information available to the players, and the utilities they receive for different outcomes. Once the model is specified it can be analyzed using Nash Equilibrium or any of the other numerous solution concepts proposed in the literature [16].

There has been extensive research on solution concepts, but less attention has been given to the problem of specifying game models. The model is typically assumed to be given as a starting point, and common knowledge to all players. This is problematic for several reasons. Games that model real world interactions are often complex, with

O. Veliz (✉) · C. Kiekintveld
University of Texas at El Paso, El Paso, USA
e-mail: osveliz@miners.utep.edu

C. Kiekintveld
e-mail: cdkiekintveld@utep.edu

Q. Bai et al. (eds.), *Multi-agent and Complex Systems*, Studies in Computational Intelligence 670, DOI 10.1007/978-981-10-2564-8_6

huge numbers of possible strategies and information states. Formal specifications of these games may be intractable to solve, even for modern supercomputers (or humans). A second problem is that the game itself may be ambiguous and difficult to formalize. Players may not have the same understanding of the possible actions, the payoffs, and the knowledge and beliefs of the other players about the game.

We argue that most game models should be considered *abstractions* that approximate more complex situations. An abstracted game model captures some–but not all–of the relevant detail about the strategic decision. In some cases, abstraction may also be used intentionally as part of the process of analyzing a complex game. For example, work on artificial agents for Texas Hold 'em poker has made extensive use of a methodology that uses abstraction to shrink the size of the game tree before applying an (approximate) equilibrium solver to compute a strategy for the game [14].

We are interested in better understanding the effect of abstraction in game-theoretic analysis. In particular, we focus on the *strategy selection problem*: how should an agent choose a strategy to play in a game, based on an abstracted model of the game? This problem has three interacting components: (1) the method for abstracting the game, (2) the method for selecting a strategy based on the abstraction, and (3) the method for mapping this strategy back to the original game. This approach has been studied fairly extensively for poker, which is a 2-player, zero-sum game. However, much less is known about how abstraction interacts with strategy selection in more general games.

The main contributions of our work are as follows. First, we specify a model of the strategy selection problem when players use asymmetric abstractions as a *meta-game*. In this model players can use different methods for abstracting the game, solving the game, and reverse-mapping the solution. We introduce a collection of specific methods for abstracting and solving games; these are intended to be representative of the most popular methods used in the literature. Finally, we present the results of extensive simulation that evaluate the candidate abstraction and solution methods on different classes of games. Our results lead to several unique observations as well as identifying solution methods that are more robust than others to error introduced by abstraction.

6.2 Abstraction Meta-Games

We first introduce a formal model that can be used to study the situation where players select strategies based on abstracted game models. Our model is based on the meta-game framework introduced by Kiekintveld et al. [11], which focused on situations where players received noisy observations of the same underlying game and had to select strategies based on these observations. The situation where players use abstractions is similar in that the players make strategy choices based on imperfect abstractions of the game. Opposing players may also use different abstractions which may cause problems for solution concepts that rely on coordination (such as Nash equilibrium).

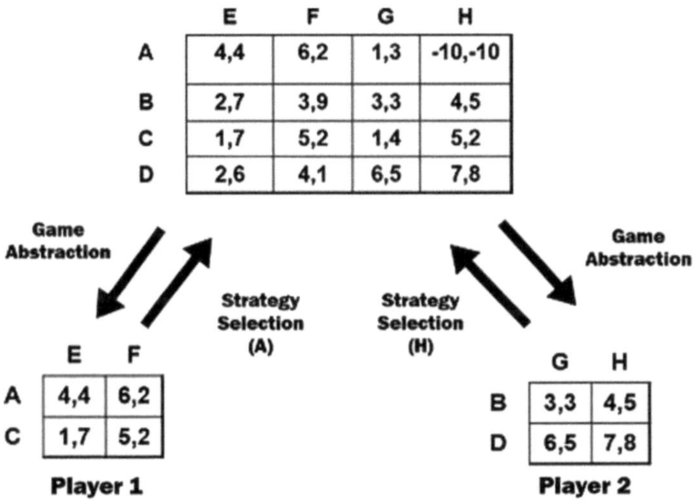

Fig. 6.1 2-players asymmetric abstractions

An example of an *abstraction meta-game* is shown in Fig. 6.1. In this example, we have two players who are playing the one-shot normal form game shown at the top of the figure; this is the *base game*. Each player has four possible actions in the original game, and the payoffs are listed in the the matrix. Each player uses a different (unspecified) abstraction method to reduce the size of the game to only two actions for each player, as shown. Now the players analyze these smaller games to select a strategy to play. Here, both of the small games can be solved using dominance to find a unique Nash equilibrium. The players select these strategies (A and H) to play. However, when these strategies are played in the base game they result in the outcome $-10, -10$, which is the worst possible payoff for both players!

In addition to illustrating the structure of the abstraction meta-game, this example shows one of the possible problems with using abstraction in game analysis. Note that we could replace the payoffs $-10, -10$ with arbitrarily small values. By playing according to a (dominant strategy) Nash equilibrium of the smaller abstracted game, the payoffs the players actually receive are less than the payoffs they expect to received by an unbounded amount. They can also be lower than the equilibrium payoffs in the original game, or even the maxmin payoffs by an arbitrary amount.

This emphasizes the problem of selecting strategies based on abstracted games. Many existing applications of automated abstraction (e.g., in Poker [8, 10, 14, 15]) still analyze the abstracted games to find an (approximate) Nash equilibrium, and play the original game using the equilibrium strategy for the abstracted game. Much of this work focuses on zero-sum games where Nash equilibrium is equivalent to the pessimistic maxmin solution concept, and has a unique value. However, for general-sum games it is clear from our example that the simple approach of analyzing abstracted games using solution concepts that assume perfect knowledge of the game model is highly problematic. Even in the case of poker, there have been pathologies shown in using abstraction [18], which has led to the exploration of alternative methods for playing games using abstraction that are not based on playing according to an approximate Nash equilibrium [7, 9].

We refer to the full model as a *meta-game*, and strategies in the meta-game as *meta-strategies*. The general strategy selection algorithms we are interested in map directly to meta-strategies in this model. We can view players as selecting meta-strategies (solution algorithms) that act on their behalf once game observations are made. Modeling solution algorithms as strategies in a larger game motivates evaluation of the algorithms with game-theoretic concepts, such as regret and equilibrium.

We now present a more formal description of abstraction meta-games. Both the base game and abstracted games are instances of normal-form games, which can be defined by a tuple $\{I, \{S_i\}, \{u_i(s)\}\}$. The set I represents the players, and the sets S_i are the pure strategies available for each player. The utility function u_i maps each outcome to a real number representing the payoff for the player if this outcome is played. We extend the model to allow mixed strategies in the standard way, using the notation σ_i to refer to a mixed strategy that is a probability distribution over the pure strategies. Payoffs for mixed strategies are defined using the expected utilities based on the pure-strategy outcomes.

A *Nash equilibrium* is a strategy profile in which every player is playing a best-response to the opponent's strategies. That is, for all players i and all strategies $s_i \in S_i, u_i(\Sigma_i, \Sigma_{-i}) \geq u_i(s_i, \Sigma_{-i})$. We also define the *regret* for a strategy profile to the maximum benefit to be the maximum gain that any player can gain by deviating to a different pure strategy. Any strategy profile with zero regret is Nash equilibrium.

An abstraction meta-game has a base game G in the normal form. This is the game that the players will actually play, and their strategy choices in this game define the payoffs they receive. The meta-game extends the description of the base game to include a description of how players select their strategies in the base game. This description has three components. The *abstraction function* (ω_i) for a player maps the base game G into a smaller abstracted game G'_i that the player will analyze. The *selection function* for a player maps from the abstracted game G'_i into a strategy for the player to play in the abstracted game, σ'_i. The *reverse mapping function* for a player maps from the strategy in the abstracted game σ'_i back to a strategy in the original game σ_i. Collectively, we refer to these three functions as a players *meta-strategy*. The meta-strategy represents a detailed description of how a player analyzes the base game to select a strategy to play.

6.3 Abstraction Methods

We define an abstraction method as a function that maps one normal-form game into a second (smaller) normal-form game. A good abstraction should be simpler than the original, but also retain as much information about the strategic interaction as possible. We identify two broad categories of abstractions that are common in the literature: *strategy elimination* and *strategy combination*.

Strategy elimination abstractions remove some strategies completely to reduce the size of the game. The well-known method of iterated elimination of dominated strategies [16] is one example of this approach. Weakening the criteria for dominance so that weakly or approximately dominated strategies are also removed leads to even smaller games at the cost of potentially removing some equilibria of the original game [3].

Abstractions methods based on strategy combination simplify games by merging multiple strategies into a single representative strategy. In the extreme case where two strategies have exactly identical payoffs for a player in every outcome merging these strategies results in a lossless abstraction. The idea can be extended to merge strategies that are only similar rather than identical, but the result is a lossy abstraction. A common approach in the literature is to use clustering algorithms to merge similar strategies into clusters. In extensive form games, clustering can also be applied to information states [8]. A different type of similarity between strategies can be used to solve a game by reducing it to sub games [4].

Our goal in this paper is not to develop novel abstraction methods, or to exhaustively evaluate the many existing techniques. We are interested in the interaction between abstractions and the solution methods applied to abstracted games. Therefore, we selected two simple abstraction methods for our initial study that are representative of the broad categories described above. We also do not try to guarantee any bounds on these general abstractions as shown in Fig. 6.1 with potentially unbounded error. It is also the case that the simulated agents may not be using an equilibrium or convergent strategy.

6.3.1 *TopN Abstraction*

The first abstraction method we consider is *TopN*. This method creates a smaller game by selecting a subset of size N of the strategies for each player to form the abstracted game. For each strategy in the game, we first calculate the expected payoff of the strategy against a uniform random opponent strategy. We then select the N strategies with the highest expected payoffs, breaking ties randomly. The abstracted game is simply the game where players are restricted to playing only the N selected strategies; the payoffs are unchanged. Since each strategy in the abstracted games is also a strategy in the original game the reverse mapping of the strategies back to the original game is trivial.

6.3.2 *K-Means Clustering Abstraction*

The second abstraction method we use is *KMeans*, which is representative of strategy combination methods. This method uses the k-means clustering algorithm to group strategies into clusters based on similarity of their payoffs. Each strategy is represented in the clustering algorithm by a vector of the payoffs for the strategy in every outcome that can result when the strategy is played. We cluster the strategies for each player separately, following the standard k-means algorithm. Each strategy is initially assigned randomly to one of the k clusters. The centroid is calculated for each cluster, and then the Euclidian distance between every strategy and every cluster centroid is calculated. Strategies are re-assigned to the closest cluster, ensuring that no cluster becomes empty. This process iterates until there are no further changes in the clusters. We run the clustering 100 times with different initial clusterings, and select the one with the smallest maximum distance between any strategy in a cluster and the centroid.

Once the strategies are clustered we create the abstracted game as follows. Each cluster maps to a pure strategy in the abstracted game (so the number of strategies is controlled by the k parameter used in k-means). The payoffs for each outcome in the abstracted game are computed by averaging the payoffs for all of the outcomes in the cluster. In other words, we assume that players will play each strategy in a given cluster with equal probability. The reverse mapping also assumes that players play strategies in the same with uniform random probability. For example, if a strategy in the abstracted game places a probability of 0.5 on playing a given pure strategy, this probability would be distributed uniformly over all of the strategies that comprise that cluster in the strategy used to play the original game.

6.4 Candidate Solution Methods

The task of our solution methods is to select a strategy to play in a game. We consider several candidate solution methods for selecting strategies in abstracted games. All of these are based on known solution concepts or simple heuristics for playing games, and they are intended to provide a diverse pool of plausible strategies to evaluate. Little is known of the interactions among these commonly used solution techniques in the presence of abstraction or even in different classes of games. Of particular interest are several solution concepts that originate in behavioral game theory and have been shown to predict human behavior better than Nash equilibrium and related concepts. We hypothesize that humans may be adopting these types of strategies in part because they are more robust to ambiguity and uncertainty.

Uniform Random (UR): Play each pure strategy with equal probability.

Best Response to Uniform Random (BRU): Play the pure-strategy best-response to UR. It is equivalent to a level-1 strategy in the cognitive hierarchies model [2].

Nash Equilibrium (MSNE): We use Gambit [5] logit solver to calculate a sample Nash equilibrium in mixed strategies and play according to this strategy.

Epsilon-Nash Equilibrium (ENE): For every pure-strategy profile, we first calculate the maximum that value (ε) that any player can gain by deviating to a different pure strategy. We select the profile with the smallest value of ε (the best approximate Nash equilibrium in pure strategies) and play the associated pure strategy.

MaxMin: Play the strategy that maximizes the worst-case payoff for the player.

Fair: This heuristic strategy focuses on outcomes that are "fair" in that there is a small difference between the payoffs for the players. For every strategy profile we calculate the difference between the payoffs and select an outcome that minimizes this difference. Ties are broken in favor of outcomes with a higher sum of payoffs, and then randomly.

Social: This strategy plays according to the outcome that maximizes the sums of the payoffs for all players. If there are ties the strategy plays a uniform random strategy over the strategies in the tied outcomes.

Quantal Response Equilibria (QRE): Quantal Response Equilibrium [13] originated in behavioral game theory. It incorporates a model of *noisy best-response* where players use a softmax decision rule instead of strict best-response. A logistic function is normally used to specify this response, and it has a parameter λ that interpolates between a uniform random strategy when $\lambda = 0$ and a best response as $\lambda \to \infty$. A QRE is defined similarly to a Nash equilibrium, except that both players play noisy best-responses to each other for some value of λ. QRE has been shown to provide a better fit for experimental data on human behavior in some games [13]. QRE has also been shown to have more robust strategy choices that NE in situations where players make choices based on noisy observations of an underlying game [11]. We compute a QRE using Gambit [5] and play a best-response to the predicted QRE strategy of the opponent.

Cognitive Hierarchies (CH): Cognitive Hierarchies [2] also originates in behavioral game theory. It models a recursive style of reasoning where level-0 agents play a uniform random strategy, level-1 agents play a best response to the level-0 agents, level-2 agents play a best response to a mixture over level 0 and 1 agents, etc. Agents at each level use a Poisson distribution, based on a parameter τ, to predict the probabilities of playing agents at lower levels, and play a best response to this mixture.

Quantal Level-k (QLK): This method combines the features of QRE and CH. It uses a recursive reasoning model identical to CH, except that in place of the best-response at each level agents play a noisy best-response using the same logit function used in QRE. It has parameters for both λ and τ. We play a best-response to the predicted strategy of the opponent, based on the specified level of reasoning.

6.5 Experimental Methodology

We run simulations using the meta-game framework from Sect. 6.2. Our testbed allows us to run tournaments with different combinations of (1) classes of base games, (2) abstraction methods, and (3) solution methods. The simulation first generates a random game from the class as the base game. Each meta-strategy is a combination of an abstraction method and a solution algorithm. We calculate the strategies selected by each meta-strategy and then play a round-robin tournament among these strategies. The payoffs are based on the strategy profile that is played in the original game.

The result of the tournament is a payoff matrix for the meta-game where each meta-strategy can be played by either player. To estimate this matrix for a class of base games we average the payoffs over a large number of sampled games. We can analyze this payoff matrix to generate several performance metrics for the meta-strategies. Average payoffs are quite biased, in that they can reward strategies that only do well against other weak strategies. Instead, we present our results using *stability* and *average regret* measures. Stability is a measure of how close the pure strategy profile where all players use the same meta-strategy is to being a pure Nash equilibrium. It is calculated by finding the maximum gain for deviating to another meta-strategy from this profile. Any value less than or equal to zero indicates that the profile is a pure equilibrium of the estimated meta-game. Average regret is the average of the regret for playing a given meta-strategy against each other meta-strategy in the estimated game.

We use three classes of randomly-generated games: general sum, zero sum, and logical games. General sum and zero sum are generated using GAMUT [6]. Players have 20 pure strategies, and payoffs are real numbers in the range [−100, 100]. Logical games have more structure and should be more amenable to abstraction. Pure strategies are based on choosing either 0 or 1 for a set of boolean variables. Payoffs are based on randomly generated oppositions with varying numbers of clauses, literals, and values. If a proposition is true for a given outcome in the game, the payoff associated with the proposition is added to the payoff for that player. The payoffs are additive over all true propositions for a given outcome. Finally, payoffs are normalized to the range [0,100]. We also could not possibly evaluate every level of abstraction so we picked several interesting abstraction levels. An abstraction level of 10 means that there are now 10 actions in the game, Top10 or 10 clusters, and 20 indicates these are the original 20 pure strategies with no abstraction.

6.6 Results

Our experiments are designed to test the relationships between abstraction methods and solution concepts across different classes of games, and with varying levels of abstraction. In total, we experiment with three different classes of games, two types of abstraction, four levels of abstraction, and more than 30 different solution methods (including multiple parameter settings for QRE, CH, and QLK). The main results span 24 different tournaments with 500 sample games played in each one.

The full data set is difficult to visualize, so we will present results focusing on selected subsets of agents to illustrate key points. We focus on measures of stability and regret here as we believe they are the most informative, but we have also looked at measures of average payoffs and exploitability. Average payoffs can be misleading for agents that exploit weak players but perform poorly against strong players. Exploitability does not differentiate well among the meta-strategies, since almost all of them are highly exploitable in the worst case.

We first present a set of results that demonstrates how we compare the different parameter settings for QRE, CH, and QLK. The purpose of this analysis is to discover which parameter settings are strongest. In later results we will include only best parameter settings for each solution method to aid in visualization.

Figures 6.2 and 6.3 show results for QRE with different values of the λ parameter. Each plot shows results for all three classes of games. There are separate plots for each type of abstraction. The x-axis shows the level of abstraction where the values are the number of actions in the abstracted game. The point on the far left corresponds to no abstraction, and more information is lost for data points further to the right. The y-axis represents the stability (ε) value for each solution method.

Lower stability values are better. In particular, any value less than or equal to 0 indicates that a solution method is a pure-strategy Nash equilibrium, and if all players were using this method none of them would benefit by deviating to any other solution method. These stabilities are calculated with respect to the full data set, meaning that players are allowed to deviate to *any* solution method, not just the ones in the figure.

In general sum and logical games, the best QRE parameter settings are closer to playing a best-response to a more random opponent than playing a best-response to an opponent closer to a Nash equilibrium. For example, the very low parameter setting of $\lambda = 0.05$ performs very well, and this setting is the closest to uniform

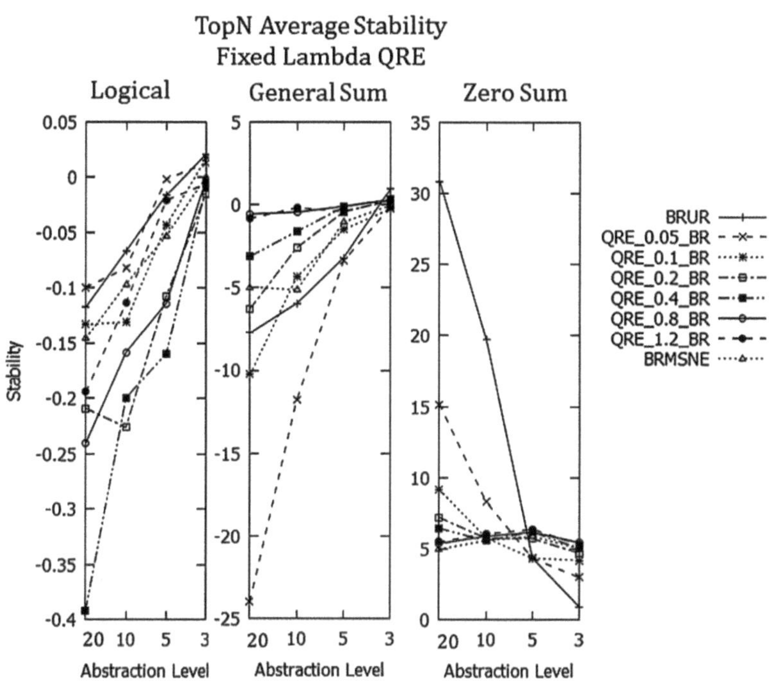

Fig. 6.2 TopN stability fixed lamda QRE

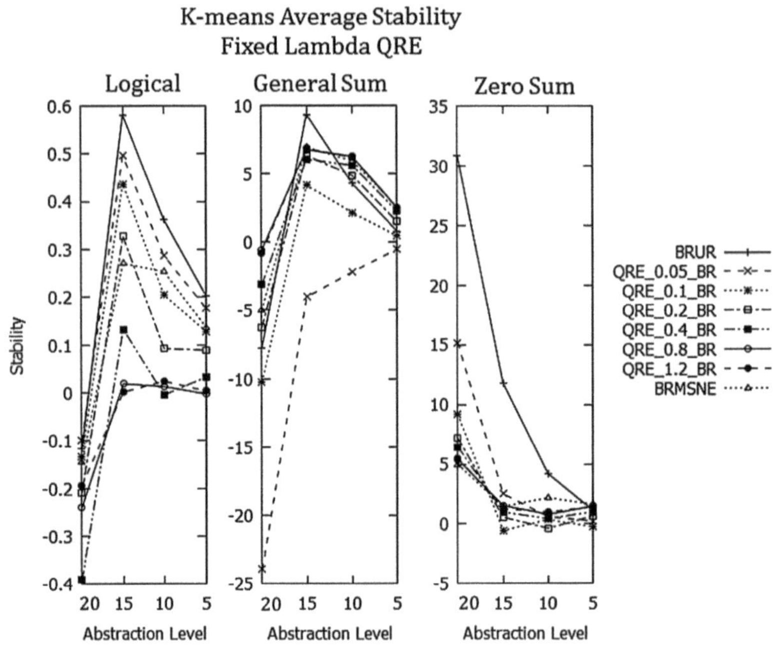

Fig. 6.3 K-means stability for fixed lambda QRE

random. Zero-sum games appear to behave quite differently where playing closer to an equilibrium strategy performs better.

We perform a similar set of experiment to determine the parameterizations of CH and QLK agents for the values of k, τ, and λ. We omit them due to space considerations, but use a similar procedure to identify the most effective parameter settings to visualize in the later plots. The values of τ considered are 1.5, 2.5, 5, 7.5, and 9, the values of k are 0, 1, 2, 3, 5, 10, and 20, and the values of λ are 0.8, 1.5, 2.0, 5.0, and 20. For QLK agents we also consider both version that play the raw QLK strategy and ones that play a best response to the strategy of the opponent for a particular level. One results to note from this analysis is that for CH agents with the same τ but different levels of reasoning there is not much difference between agents with mid-level k versus high level k. The main variations between these agents occur at lower levels. However, higher levels of k typically have better, more robust performance.

We now turn to analysis of the complete set of agents. The following results are from a fully symmetric tournament of 500 games for each class of games and 30 agents which include the best parameter settings for QRE, CH, QLK, and the other main solution concepts. We selected the best parameter settings for each of the QRE, CH, and QLK variants based on the results of the previous analysis to visualize in these results along with the other agents.

Figures 6.4, 6.5 and 6.6 show the average stability and regret for the different solution methods (excluding the lines for all but the best QRE, CH, and QRE settings) when using the KMeans and TopN abstractions in different classes of games. We begin by noting some general trends in the results. The results for logical games in Fig. 6.6 shows how all of the stability and regret values are very low compared to both general-sum and zero-sum games. There is also very little variation among each of the strategies. This indicates that these games are much easier to play well in, even when they are abstracted. In most cases, agents are more stable for cases without any abstraction, and less stable but converging to the same range as abstraction increases. However, regrets are very high for cases without abstraction and lower for cases with abstraction. Zero-sum games behave somewhat differently; the overall stability is worse, and the stability of many of the agents actually improves (surprisingly) as abstraction increases.

There is no solution concept that is clearly the best across all of the settings tested. Instead, the best strategy depends heavily on both the class of games and the type of abstraction being used. The UR and Fair agents do poorly in almost all cases. The ENE and Social agents perform surprisingly well, particularly when using the TopN abstraction and in the general-sum and logical games. These agents focus on coordinating on high payoffs, so it is interesting that even these simple strategies are able to do so well even in cases with abstraction error. For the cases with KMeans abstraction and for zero-sum games, approaches close to equilibrium solvers perform better particularly MSNE and the QRE and QLK.

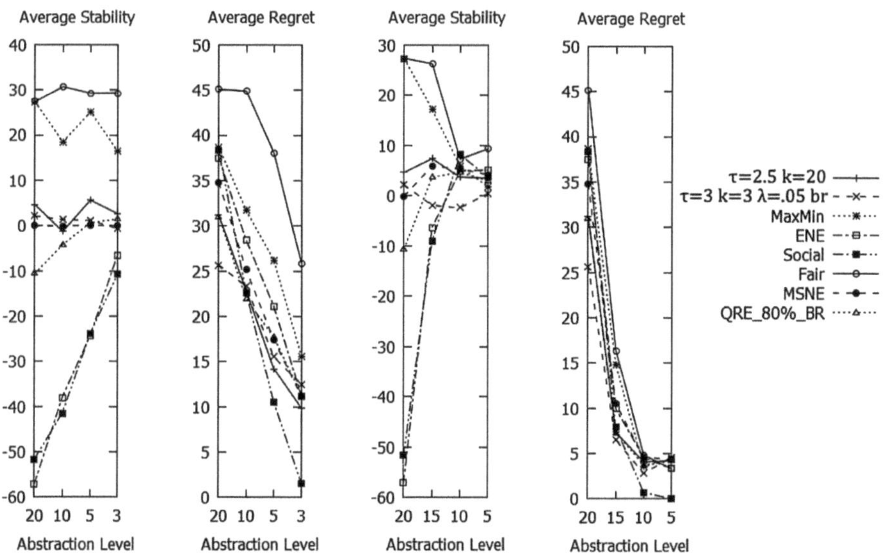

(a) TopN (b) TopN (c) K-means (d) K-means

Fig. 6.4 Average stability and regrets for general sum games

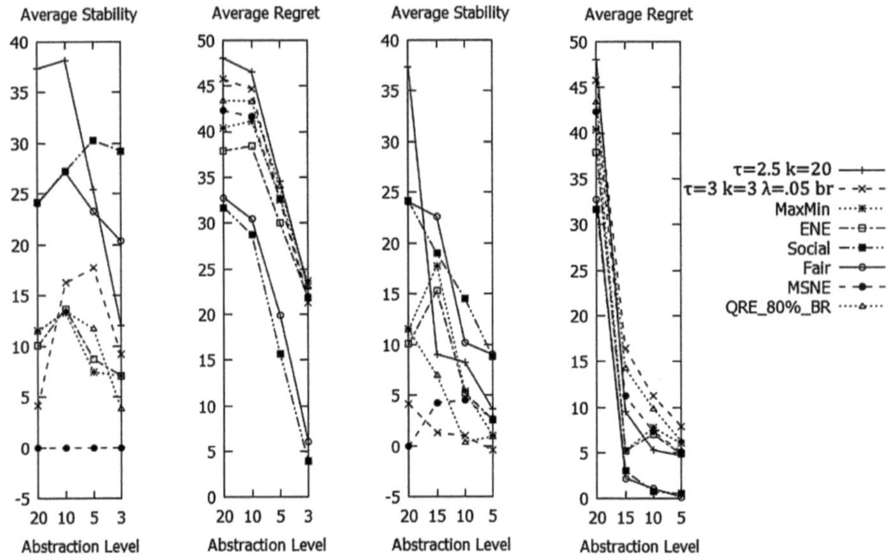

(a) TopN (b) TopN (c) K-means (d) K-means

Fig. 6.5 Average stability and regrets for zero-sum games

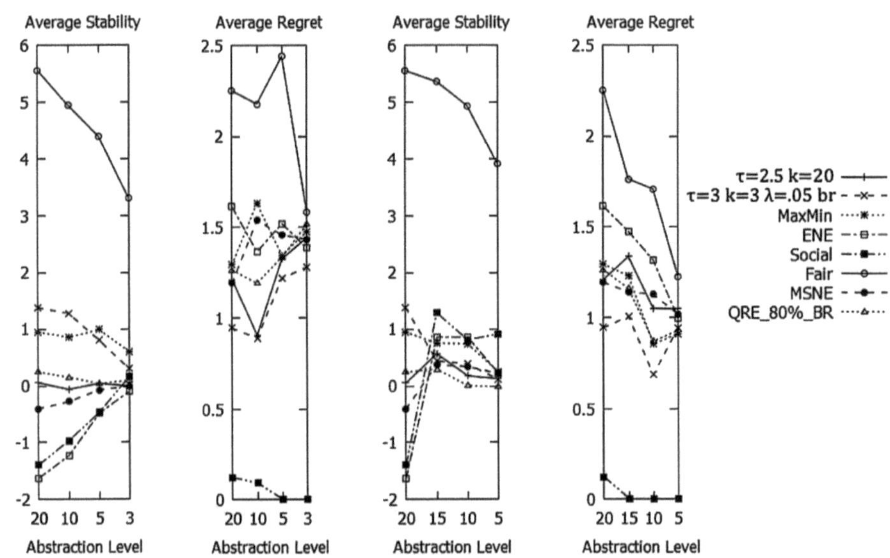

(a) TopN (b) TopN (c) K-means (d) K-means

Fig. 6.6 Average stability and regrets for logical games

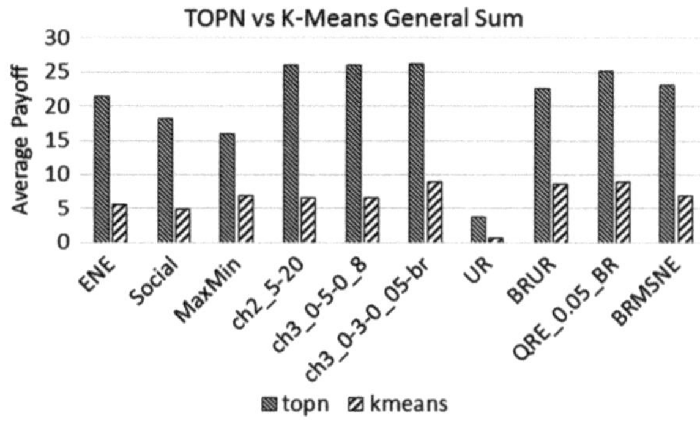

Fig. 6.7 TopN versus Kmeans in general sum—abstraction level 5. The third CH is QLK

While this pattern of results fails to identify a clear winner among the common solution methods for how to select strategies in abstracted games, the results are still quite interesting. They clearly show that both the game class and abstraction method interact strongly with the applied solution concept, and simply finding a NE is not the best approach. There is also an opportunity to develop significantly improved solution concepts that account for these factors unlike current practices.

The final experiment we present directly compare the performance of the two different types of abstractions. In this experiment, players can use either abstraction, which leads to even more asymmetry in the abstractions players are using.

In Fig. 6.7 we show the average payoffs for the different solution methods in a tournament that combines agents using both abstractions. We only use the best parameter settings for the QRE, CH, and QLK agents. This tournament was run on general-sum games, using the highest level of abstraction (from a game size of 20 down to a game size of 5). For each solution method we show the average payoffs achieved when using the two abstraction methods side by side. Interestingly, the TopN abstraction method outperforms KMeans in combination with *every one* of the solution methods, often quite substantially. The TopN abstraction plays a role in coordinating agents on high-payoff outcomes independent the solution algorithm.

6.7 Conclusion

Our results demonstrate that using abstraction to solve games is a complex endeavor, and the type of abstraction, the solution methods used to analyze the abstracted games, and the class of games all have a strong influence on the results. Many of the strongest results using abstraction to analyze large games (e.g., Poker) have focused on zero-sum games. One of the most interesting observations from our results is that abstraction often works very differently in zero-sum games than it does general-sum games or the more structured logical games. In particular, solution methods based on finding Nash equilibrium seem to work much better in zero-sum games than they do in the other classes of games in our experiments. Another important observation from our experiments is that Nash equilibrium often does not perform well in cases where abstraction is used as part of the solution process. It is still effective when the games are zero-sum, but in the other cases it was not robust to the introduction of error based on game abstraction. The two solution methods that were the most robust to abstraction error across the many different settings were BRU and QRE. Of the two, QRE was more consistent and had stronger performance in most cases, but it was not dominant in all situations.

We also found that the specific method used for generating abstractions has a strong impact on the results. One very interesting result was that TopN was sometimes able to *increase* the payoffs for the agents in comparison to the case without any abstraction. However, TopN is a symmetric abstraction when both players use it, while KMeans is asymmetric. Some methods like the social agent performed much better when using the symmetric TopN abstraction than when using the asymmetric KMeans abstraction. This kind of interaction is very important to understand in greater depth if we are to make effective use of abstraction as part of game-theoretic analysis. Our model of abstraction meta-games provides a formal model for studying this type of interaction, and our simulations have resulted in several interesting observations that provoke many additional questions about the use of abstraction in solving games.

Future work will include other types of abstraction and tests on even larger games. We plan to also develop new robust solution techniques that can perform well in the presence of abstraction. It is also important to test on games with more realism like a simplified version of poker or a security game to examine how general abstraction and robust strategy selection will perform in these kinds of games.

Acknowledgments This material is based upon work supported by the National Science Foundation under Grant No. IIS-1253950.

References

1. Alpcan, T., Basar, T.: Network Security: A Decision and Game Theoretic Approach. Cambridge University Press (2011)
2. Camerer, C.F., Ho, T.H., Chong, J.K.: A cognitive hierarchy model of games. Q. J. Econ. **119**(3), 861–898 (2004)
3. Cheng, S., Wellman, M.: Iterated weaker-than-weak dominance. In: IJCAI, pp. 1233–1238 (2007)
4. Conitzer, V., Sandholm, T.: A technique for reducing normal-form games to compute a nash equilibrium. In: Proceedings of the 5th International Joint Conference on Autonomous Agents and Multi Agent Systems (AAMAS-06), pp. 537–544 (2006)
5. Gambit: Gambit. http://gambit.sourceforge.net/ (2014). Accessed 01 Aug 2014
6. GAMUT: Gamut. http://gamut.stanford.edu/ (2014). Accessed 01 Aug 2014
7. Ganzfried, S., Sandholm, T., Waugh, K.: Strategy purification and thresholding: effective non-equilibrium approaches for playing large games. In: Proceedings of the 11th International Conference on Autonomous Agents and Multiagent Systems. AAMAS'12, vol. 2, pp. 871–878 (2012)

8. Gilpin, A., Sandholm, T.: Expectation-based versus potential-aware automated abstraction in imperfect information games: an experimental comparison using poker. In: Proceedings of the Twenty-Third AAAI Conference on Artificial Intelligence (2008), pp. 1454–1457 (2008)
9. Johanson, M., Bard, N., Burch, N., Bowling, M.: Finding optimal abstract strategies in extensive-form games. In: AAAI (2012)
10. Johanson, M., Burch, N.: Evaluating state-space abstractions in extensive-form games. In: Proceedings of the 12th International Conference on Autonomous Agents and Multiagent Systems (AAMAS), pp. 6–10 (2013)
11. Kiekintveld, C., Wellman, M.: Selecting strategies using empirical game models: an experimental analysis of meta-strategies. In: Proceedings of the 7th International Joint Conference on Autonomous Agents and Multiagent Systems, AAMAS'08, vol. 2, pp. 1095–1101 (2008)
12. Krishna, V.: Auction Theory. Academic Press (2002)
13. McKelvey, R., Palfrey, T.: Quantal response equilibria for normal form games. Games Econ. Behav. **10**(1), 6–38 (1995)
14. Sandholm, T.: The state of solving large incomplete-information games, and application to Poker. AI Mag. Spec. Issue Algorithmic Game Theor. 13–32 (2010)
15. Sandholm, T., Singh, S.: Lossy stochastic game abstraction with bounds. In: Proceedings of the 13th ACM Conference on Electronic Commerce. EC'12, vol. 1, pp. 880–897 (2012)
16. Shoham, Y., Leyton-Brown, K.: Multiagent Systems: Algorithmic, Game-Theoretic, and Logical Foundations. Cambridge University Press (2009)
17. Tambe, M.: Security and Game Theory: Algorithms, Deployed Systems, Lessons Learned. Cambridge University Press (2011)
18. Waugh, K., Schnizlein, D., Bowling, M., Szafron, D.: Abstraction pathologies in extensive games. In: Proceedings of the 8th International Conference on Autonomous Agents and Multiagent Systems (AAMAS), pp. 781–788 (2009)

Chapter 7
Adaptive Forwarder Selection for Distributed Wireless Sensor Networks

Nor Azimah Khalid and Quan Bai

Abstract Wireless Sensor Network has emerged as a promising networking technique for various applications. Due to its specific characteristics, such as non-rechargeable, low-power multi-functional sensor nodes, limited sensing, computation and communication capabilities, it is challenging to build networking protocols for Wireless Sensor Networks. In this chapter, the focus is on addressing the routing issue with regards to energy efficiency and network lifetime. An adaptive and self-organized routing protocol for distributed and decentralized network, called Distributed Adaptive Forwarder Selection, is proposed. Multiple factors, involving cross layers were used for selecting the adequate forwarders for packets. The proposed approach is suitable for dynamic environments as there is no fixed topology or static role assignment for nodes in the WSN. In addition, the approach can allow sensor nodes to make flexible decisions based on their current capabilities and states. We have performed simulations of the proposed protocol and compared with two existing routing protocols in terms of node lifetime, average energy consumption and average residual energy. The results show that the proposed protocol performed better than some well known routing protocols such as LEACH and MOECS.

Keywords Distributed wireless sensor networks · Forwarder selection · Reinforcement learning

7.1 Introduction

In general, a Wireless Sensor Network (WSN) is a wireless network which consists of large numbers (hundreds to thousands) of irreplaceable and low-power multi-functional sensor nodes, operating in an unattended environment with limited sensing, computation and communication capabilities [1] used in a wide range of

---

N.A. Khalid (✉) · Q. Bai
Auckland University of Technology, Auckland, New Zealand
e-mail: nkhalid@aut.ac.nz

Q. Bai
e-mail: quan.bai@aut.ac.nz

Q. Bai et al. (eds.), *Multi-agent and Complex Systems*, Studies in Computational Intelligence 670, DOI 10.1007/978-981-10-2564-8_7

applications. Physical resources, such as memory, communication bandwidth and energy, can greatly limit the capability of sensor nodes and the performance of the whole WSN system [2]. A number of previous works have focused on these constraints for designing the communication and information processing elements for wireless sensor networks.

Communication process has been identified as highly resource consuming especially when the process is not well managed [3, 4]. There are two elements of communication for wireless sensor networks application: routing mechanism and media access control (MAC) protocol. In previous studies, several works for energy perspective issues have been briefly described on both elements. This involves reducing number of transmissions and distance of transmission via clustering or scheduling mechanisms at specific network layers.

For large scale networks, decentralized architectures are more appropriate as high transmission cost and delay might involve, especially if the central controller is located far away. Furthermore, in a centralized architecture, if the central node fails, then the entire network will collapse. On the other hand, decentralized control architecture are more reliable for large networks and can provide better collection of data and backup, in case of failure of the central node. However, decentralized approach is very challenging in terms of topology establishment and re-establishment especially in inaccessible applications such as battlefield and disaster management. These non replenish nodes need to self-organize themselves to sustain longer. This requires nodes in decentralized scheme to adapt accordingly with dynamic changes of environment (i.e., the network topology) and self-configure themselves without human intervention.

Shortest path algorithms based on either hop count or energy consumption are typically employed in routing protocols of ad hoc networks to achieve high energy efficiency [5]. However, relying on these parameters might cause hot spot scenario, i.e., sensor nodes that are frequently used might get depleted. In such case, a more reliable nodes should be selected. This leads to the need of having a more adaptive parameters to be considered during path selection. Furthermore, node preferences might be different, i.e., one node will choose a path which is nearer to it, which could be far to other nodes.

In this chapter, we focus on node selections for packet relaying (i.e., forwarder selection), and propose an adaptive forwarder selection approach for distributed WSNs. The proposed approach is called Distributed Adaptive Forwarder Selection (DAFS). In DAFS, suitable forwarders are selected in three phases, i.e., Eligibility Determination, Forwarder Selection and Receiver Acceptance. Multi-criterion parameters including energy, distance and buffer size, are considered in the approach. We claim that DAFS is an adaptive approach and suitable for dynamic environments, where nodes actions are determined based on their current capabilities and state of environment. In addition, it is suitable for largely distributed networks, where only decentralized approach is feasible.

The rest of the chapter is organized as follows. Some related work is introduced in Sect. 7.2. In Sect. 7.3, we describe the targeted problems and give some formal definitions. Section 7.4 introduces the interaction protocol applied in the proposed approach. Section 7.5 explains the forwarder selection processes in DAFS. The experimental results have been presented in Sect. 7.6. The chapter is finally concluded in Sect. 7.7.

7.2 Related Work

In this section, we briefly review related work on energy-efficient routing protocols and learning-based protocols. Most of the previous studies have shown that clustering approach can reduce the energy expenditure when merging all the information from the nodes into one cluster head which is responsible to process it and deliver it to the sink or base station. The limitation of sensor nodes usage for processing information has given a better energy consumption management which results to the sensor nodes lifetime to increase.

One of the well-known cluster-based protocols is Low Energy Adaptive Clustering Hierarchy (LEACH) [6]. In LEACH, data collection is perform periodically, which involves two phases, i.e., Cluster Head (CH) selection and cluster formation. The selection of CH in LEACH is based on closest distance. Each CH creates Time Division Multiple Access (TDMA) schedule for member nodes. CH also select code division multiple access (CDMA) to reduce inter-cluster interference. Members will collect information and use their allocated TDMA slots to transmit their collected data to CH. In [7], an energy efficient cluster formation algorithm (MOECS), was proposed based on a multi-criterion optimization technique. The selection of cluster heads is restricted to certain optimal value (i.e., optimal radius and distance from normal nodes).

Learning-based approach is commonly used in distributed systems [8]. Distributed Independent Reinforcement Learning (DIRL), is based on independent learning, i.e., each agent can autonomously and dynamically self-configure in order to maximize its own reward [9]. A reward-based dynamic approach based on two tier reinforcement learning scheme (micro learning and macro learning) were proposed in [10]. In their approach, an individual node was able to self-schedule its task using its local information and through learning [11].

Q-learning is a model-free RL technique, based on agents taking actions and receiving rewards from the environment in response to those actions [12]. In [13], Dimarogonas and Johansson proposed a combinatorial reverse auction that operates in two phases using RL and some economic models for energy optimization in sensor networks. The Q value was represented as an estimate cost of the route through neighbor comprised of hop count (account for energy efficiency) and minimum battery level among nodes. In [14], the potential of using energy aware metrics in RL based routing algorithms for WSN was studied, combining energy aware metrics with load balancing metrics. In [5], a machine-learning-based routing protocol for energy-efficient and lifetime-extended for UWSN, i.e., QELAR, was proposed. In QELAR, residual energy of each node and energy distribution among a group of nodes were used in it's lifetime-aware reward function, for calculating the Q-value (in selecting forwarder for packets). In [15], role-free clustering assignment was combined with learning dynamic network properties such as battery reserves. Less energy was consumed by using machine learning to enable nodes to independently decide whether or not to act as a cluster head on a per-packet basis in comparison to a traditional approach.

7.3 Problem Description and Definitions

In this chapter, we propose an adaptive, energy efficient and lifetime-aware forwarder selection approach, based on Q-learning technique. Using an action-value function (Q-value), which gives the expected reward of taking an action in a given state, the distributed learning agent is able to make a decision automatically. The proposed approach has the following features:

- Dynamic Network: In largely distributed network, link quality is not guaranteed. Link failure due to node's energy depletion, causes topology changes. Using Q-learning algorithm, selection of alternative link is possible as node selects next best forwarder based on current situations.
- Adaptive: We define our node role as forwarder, receiver and normal nodes. Node will decide on its role based on its present capabilities. It is adaptive to available resources i.e., a receiver can accept packets that they are capable to process, i.e., accept more when less busy. A node can be forwarder at a time but not at the other time if there is other more capable node to forward the packet etc.).
- General Framework: Q-learning behavior is determines by its reward function. We proposed a flexible and dynamic approach for the nodes to react based on its present capabilities.
- Load Balancing: Less energy is consumes when choosing a path based on shortest path. However, this may cause the link failure as choosing the same node to forward packets could drain its energy faster. We consider multiple parameters to allow alternative path selection.

7.3.1 Definitions

The network in our model is consider as a complex system comprising a number of adaptive sensor nodes, called agents.

Definition 1 A WSN is defines as connected undirected weighted graph $G = (V, E)$, where V is group in the network comprises of agents, i.e., $V = a_0, a_1, a_2, \ldots a_n$. $E = e_1, e_2, \ldots e_m$ is a set of edge in group. The edge, $e_k = (a_i, a_j)$ denotes the communication links between sensor a_i and sensor a_j (they are in each other's radio transmission range).

Definition 2 An agent a_i is defined as $a_i \in (R_i, Act_i, C_i, RWD_i)$. R_i is a_i's Role; Act_i is the action that a_i takes; RWD_i is the Reward that a_i gains (see Definitions 4); and $C_i = (EInit_i, ERes_i, ENeigh_i, Dist, BF_i, BFNeigh_i)$ is the capability of a_i, where $EInit_i$ is a_i's initial energy, $ERes_i$ is a_i's residual energy, $ENeigh_i$ is a_i's immediate neighbors' energy, Dist is distance between agents. BF_i is current buffer size of a_i, $BFNeigh_i$ is current buffer size of a_i's neighbors.

There are three types of Role, i.e., R in the proposed model, which are forwarder, receiver and normal node. In dynamic environment such as WSN, it is not practical to select a node as forwarder permanently, as it will cause the node to die faster. We propose a more distributed approach which allow flexibility in being forwarder, based on current capabilities. A forwarder can be a normal node at other time when its energy has degraded or it is currently processing many tasks.

Definition 3 An action set Act_i is defined as: $Act_i = (act_1, \ldots act_i \ldots act_n)$, where act_i is a possible action that a_i can perform. As a forwarder, an agent can take the following two actions:

- Forward packets received from one agent to another agent.
- Discard packets if no Forwarder is identified.

As a Receiver, there are three possible actions:

- Accept packets based on current capabilities.
- Reject packets if buffer is full (currently busy).

Definition 4 Reward function RWD, represents expected reward received by agents when transiting from one agent to another agent. The goal of our algorithm is to get the packet delivered from one agent to another agent, with maximum reward, i.e., minimum cost. The reward function is described in Eqs. 7.2–7.5.

7.4 System Framework and Interaction Protocol

There are three modules in the proposed approach, which are Eligibility Determination Module, Dynamic Forwarder Selection Module and Receiver Acceptance Module. Algorithm 1 shows the steps involved. When an agent has packet to transfer, they will compare among eligible neighbors, which one is the most capable (i.e., the one having the highest Q value). If no agent can accept the packet, after timeout, it will drop the packet. Among Eligible agents, once they receive packets, they will decides whether to accept or reject the packets. The amount of packets it will accept is depending on its current capabilities i.e., accept packets that they are able to process.

Algorithm 1: Forwarder Selection.

```
begin
  for each step of episode do
    Prior to any decision, individual agent will share its information EResi, Dist, BFi
    and Ri with its neighbors;
    if current BF < maximum BF AND ERes > Minimum Energy
    then
      Decrease TIMEOUT ——(as it causes delay)
      Set Status as Eligible Forwarder
      Accept packet according to available BF
    ELSE
    if BF > maximum BF OR FULL OR ERes < Minimum Energy
    then
      Set Status as Not Eligible
      Choose Forwarder that give Qmax
```

7.4.1 *Eligibility Determination*

In the Eligibility Determination module, an agent decides whether to be a forwarder or not, based on its current capabilities (see Definition 2). The congestion or queue between receives or transmits will determine agent eligibility at the local level. A fully occupied buffer indicates agent is not capable to process any information at that particular time. The residual energy indicates eligibility at higher level, towards wider context i.e., network layer. Agent that decided to be Forwarder will inform it's neighbors about it's decision. In Dynamic Forwarder Selection, agent having more than one potential forwarder will select the best forwarder to forward packets based on Q value. Forwarder having the highest Q_{max} will be chosen. Q value is explained in Sect. 7.4.2. When forwarder receives packets from neighbors, it will process the received packets, according to its current capabilities.

7.4.2 *Forwarder Selection*

To assist agents to select suitable forwarders, we use Q-learning approach where using this approach, agent tends to select forwarder that gives maximum Q-value. In the proposed approach, both successful and failure transmissions contribute to the calculation of the Q-values. Furthermore, the approach not only concerns on selecting the best forwarders but also allows forwarders to negotiate as they wish, namely, a two directional selection. The expected reward that can be received by taking an action at time t and the state at time t is denotes in Eq. 7.1:

$$Q(s_t, Act_t) = RWD_{total} + \gamma \Sigma P_{s_t,s_{t+1}}^{Act_t} max\, Q(s_{t+1}, a) \tag{7.1}$$

In Eq. 7.1, $Q(s_t, Act_t)$ is the expected reward that an agent can receive by taking an action a_t at the state s_t. RWD_{total} is the total reward gained by the agent, which can be calculated by using Eq. 7.5. γ ($\gamma \in [0, 1]$) is the discount factor, which determines how important the future rewards are. When γ is set to 0, the system only considers the current reward and it acts similarly to a greedy algorithm. When γ is set to 1, the system will strive for a long-term high reward. The typical value of γ is within [0.5, 0.99]. Each forwarding action may succeed or fail. $P^{Act_t}_{s_t,s_{t+1}}$ is the success rate of taking action Act_t when s_t choosing s_{t+1} as the next forwarder. On the other hand, the failure rate is, $1 - P^{Act_t}_{s_t,s_{t+1}}$. $max Q(s_{t+1}, a)$ in Eq. 7.1 denotes the optimal value when taking an action, a. In this paper, we only consider the current reward and for such case, the second part of Eq. 7.1 is omitted.

As explained in Sect. 7.4.1, agent's capabilities are evaluated when determining Eligibility as forwarder. It is also used as input in reward functions, which is then applied in Q value calculation. We defined two reward functions, as in [5], comprises $RWD_{success}$ as in Eq. 7.2 and RWD_{fail} as in Eq. 7.4. If the packet forwarding attempt from a_i to a_j is successful, the reward function is shown in Eq. 7.2.

$$RWD_{success} = -g - \alpha(c(a_i) + c(a_j)) \tag{7.2}$$

In Eq. 7.2, g is the constant cost when a_i tries to forward a packet. As forwarding packet consumes energy and bandwidth, the farther an immediate node is from destination node, the more negative reward it would receive. Thus, agent will use a shorter path to reduce this cost. The weight of g is set to be 1. $c(a_i)$ and $c(a_j)$ are cost functions of residual energy of a_i and a_j respectively, which can be calculated by using Eq. 7.3 and α is the weight and is set to be 0.5. By definition, $c(a_i)$ is in the range of [0, 1], to balance the parameters in Eq. 7.2.

$$c(a_i) = 1 - ERes_i/EInit_i, \tag{7.3}$$

where $ERes_i$ is the residual energy of a_i and $EIniti$ is a_i's initial energy (refer to Definition 2).

On the other hand, if the forwarding attempt from a_i to a_j fails, the reward function is defined as the equation below.

$$RWD_{fail} = -g - \beta c(a_i), \tag{7.4}$$

where β is weight for the cost function that can be tuned. The value of β can be set to 0.5.

Based on Eqs. 7.2 and 7.4, the total reward gained by a_i (i.e., RWD_{total}) can be calculated by using Eq. 7.5.

$$RWD_{total} = RWD_{success} + RWD_{fail} \tag{7.5}$$

RWD_{total} is used in Q value calculation (Eq. 7.1) above. The far an agent from other agent is, the more energy is consumes for transmission. Thus, it will choose forwarder that is nearer to it.

7.4.3 Receiver Acceptance

In the Receiver Acceptance module, upon receiving a packet, agent will check its current processing task. It will accept packet according to its current capabilities, i.e., if it is currently processing certain task but still have available buffer, it will accept an amount of packets based on its remaining buffer.

7.5 The Distributed Adaptive Forwarder Selection (DAFS)

Many energy efficient and lifetime-aware approaches proposed solutions either at Physical layer, MAC layer, Network layer, Transport or Application layer. Even though such solutions can improve network performances in terms of network lifetime, energy efficiency, power consumptions etc., both analytical studies and experimental works in WSN highlight the important interactions between different layers of the network stack [3]. In this research, we consider multi-variables parameters involving Network layer, MAC layer and as well distance between nodes. In this section, we will elaborate on those parameters, which are used in our reward functions.

7.5.1 Multi-variables Parameters

In this research, agent capabilities are determined by energy, buffer size and distance. For most applications, a wireless sensor node is not replenish. Therefore, there is strong dependence on battery lifetime. Similar to traditional network layer, data transmission is linked to data communication area, which relates to certain layer; the link layer or MAC layer, Network layer (routing protocols) and transport layer (transport protocol).

7.5.1.1 Communication Energy

The main task of sensor node is to detect events, perform local processing and transmit the data. Power consumption can be divided into sensing, communication and data processing. In decentralized network, nodes may need to know its neighbors' latest state. However, in such network, continuous updates will require a lot of energy. We minimize such energy consumption by allowing only effected nodes to update and updates will only be sent if there is changes (i.e., if its energy is depleted an reaching a threshold value or if there is topology change, such as a new node joining the network). Hence, our concern is on communication energy as sensor node expends the maximum during this phase (transmitting and receiving data). The energy model in [7] is adopted where the amount of energy consumed for transmission, i.e., E_{TX}, of an ℓ-bit message over a distance d is given by:

$$E_{TX} = \ell \times EElect + \ell \times \varepsilon_{fs} \times d^2, \tag{7.6}$$

where ℓ is the length of message (4000 bits), $EElect$ is the base energy required to run the transmitter or receiver circuitry (50 nJ/bits) and ε_{fs} is the energy consumed in an amplifier (10 pJ/bit/m^2). The energy expended in receiving an ℓ-bit message, i.e., E_{RX} is given by:

$$E_{RX} = \ell \times EElect \tag{7.7}$$

7.5.1.2 Local Congestion Control—MAC Layer Solutions

The second issue considered is concerning local congestion, by limiting the traffic that an agent can relay. An agent may participate in the communication if it can relay the packet which is based on its communication activity. For this reason, buffer size is considered as another important factor in the proposed model, i.e., when packets arrive, they have to be processed and transmitted. If packets arrive faster than the agent can process them, the agent puts them into the buffer until it can get around to transmit them. The maximum queuing delay is proportional to buffer size. The longer the line of packets waiting to be transmitted, the longer the average waiting time is. The queue of packets waiting to be sent also introduces a potential cause of packet loss. Since the agent has a finite amount of buffer memory to hold the queue, an agent which receives packets at too high rate may experience a full queue where the agent has to simply discard excess packets.

7.5.1.3 Distance

In some cases, agents may be located far away from each other or from the Sink. Direct communication or peer-to-peer communication between nodes, especially in large distributed area is impossible, as it causes higher transmission cost and deplete faster. Thus, we consider distance as another important parameter. For example, if there are two Forwarders that is within agent's proximity, where forwarder A having more energy and less buffer, the agent might choose forwarder B, which has less energy and buffer compared to forwarder A but is nearer to it, taking into account, the significant energy consumption for longer distance communication.

7.6 Simulation Results

In this section, we evaluate DAFS by comparing it with two cluster-based approaches, i.e., LEACH and MOECS. The simulations were conducted using C++ platform. Two metrics were used to measure the performance of different protocols: first node death time and average residual energy. The first metric needs to be maximized, while second metric needs to be minimized. First node death time is the time when the battery of the first sensor node is depleted. Each sensor node has the goal of maximizing its own packet delivery to destination (that is to avoid packet loss by sending only to forwarder that is the most capable). Table 7.1 provides the common simulation parameters, which is also used in our experiments. Network lifetime is the most important performance metric for WSNs. Using this metric, DAFS, LEACH and MOECS protocols were evaluated.

The nodes in each simulation are distributed in a 100×100 m^2 region, where the location of nodes are selected randomly and that no two points have the same location. The Sink is given a fixed location. All the nodes are homogeneous and have the same capability.

Figure 7.1a shows the results of the first node death (round number) for two different network sizes. The first node death for network size 200 nodes, occurs at 710 rounds in LEACH, at 920 rounds in MOECS and at 3940 rounds in DAFS. While for network size 500 nodes, the first death round occurs at 730 rounds in LEACH, at 980 rounds in MOECS and at 3472 rounds in DAFS. This might be due to communication involves during clustering phase in LEACH and MOECS. In addition, as more criteria are considered in DAFS, i.e., including nodes buffer size allows nodes to choose other alternative forwarder.

In DAFS, multiple parameters that influence energy consumption were included. These parameters include communication cost from sensor node to the forwarder, communication cost from forwarder to the Sink, and the forwarder's residual energy, which help sensor nodes achieve balanced energy dissipation in the system.

Table 7.1 Simulation parameters

Number of nodes	100–500
Deployment area	100×100 m
Data packet size	500 bytes
Control packet size	25 bytes
$EElect$	50 nJ/bit
ε_{fs}	10 pJ/bit/m^2
ε_{mp}	0.0013 J/bit/m^4
Initial energy for sensor node	0.5 J
Network topology	Random

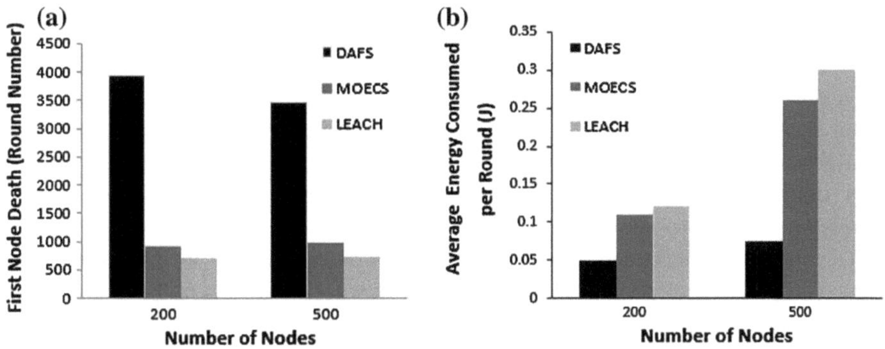

Fig. 7.1 **a** First node death in DAFS, LEACH and MOECS. **b** Average energy consumed per round in DAFS, LEACH and MOECS

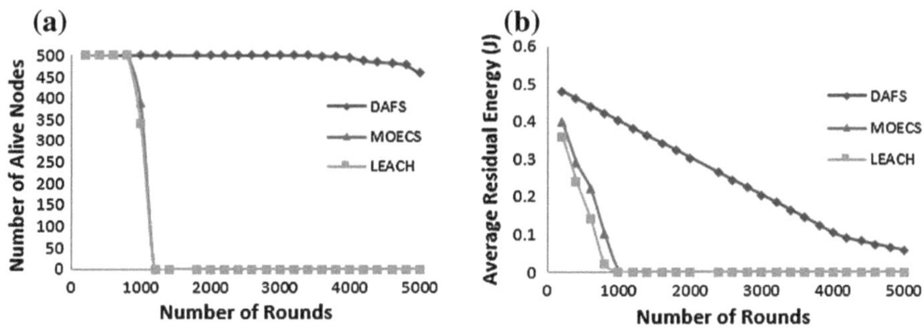

Fig. 7.2 **a** Number of alive nodes for 5000 rounds. **b** Average residual energy in DAFS, LEACH and MOECS

Figure 7.1b depicts the results for average energy consumed per round for two different network sizes using random topology which shows that our DAFS approach performs better than the other two. In addition to the balanced energy dissipation behaviors, such as distance, helps DAFS achieves minimum energy consumption compared to LEACH due to MOECS.

Figure 7.2a shows number of alive nodes in the network after 5000 rounds where nodes in DAFS survives much longer compared to the other two. Figure 7.2b illustrates results for the random topology where y-axis indicates the average residual energy and x-axis denotes the number of rounds. The residual energy of the system can also provides estimation of the network life. It can be observed that the mean residual energy of the system in the case of DAFS is higher than that of the other protocols. Hence, the network life under DAFS is enhanced compared to LEACH and MOECS. Unlike these cluster-based approaches (LEACH and MOECS), our approach did not involve cluster formation phases and is a distributed approach, as selection of forwarder is based on learning i.e., the Q_{max} value.

7.7 Conclusion and Future Work

As a resource constraint node, the use of sensor node in large scale network has some challenges in terms of energy efficiency and decentralized approach. These challenges can be overcome by ensuring energy is not use unnecessarily in transmission (multiple redundant packet, frequent use of same nodes etc.) Thus, the selection of relay node, i.e., forwarder, is crucial.

Decentralized architectures are more appropriate in many WSN applications. However, without the present of central controller, node needs to make its own decision based on limited information. In this paper, we consider multi-criteria parameters in forwarder selection and assists nodes decision by using a distributed learning-based approach. Our solution is adaptive as it is based on agent's current capabilities, that are changing dynamically when it gets depleted etc. With this technique, it is possible to consider multiple individual metrics for forwarder selection which is critical for well balanced energy dissipation of the system.

Simulation results demonstrate that DAFS achieves significant energy savings and enhances network lifetime when compared to LEACH and MOECS protocols. Multiple parameters involved in forwarder selection process for DAFS help to dissipate energy at a much more balanced rate as compared to other protocols and also it shows that the ability of DAFS to scale both from the network deployment area and node density which makes it a viable energy efficient schemes for WSNs.

References

1. Akkaya, K., Younis, M.: A survey on routing protocols for wireless sensor networks. Ad hoc Netw. **3**(3), 325–349 (2005)
2. Karl, H., Willig, A.: Protocols and Architectures for Wireless Sensor Networks. Wiley (2007)
3. Akyildiz, I., Vuran, M..C.: Wireless Sensor Networks. Wiley, New York (2010)
4. Anastasi, G., Conti, M., Di Francesco, M., Passarella, Andrea: Energy conservation in wireless sensor networks: a survey. Ad hoc Netw. **7**(3), 537–568 (2009)
5. Tiansi, H., Fei, Y.: Qelar: a machine-learning-based adaptive routing protocol for energy-efficient and lifetime-extended underwater sensor networks. IEEE Trans. Mob. Comput. **9**(6), 796–809 (2010)
6. Bsoul, M., Al-Khasawneh, A., Abdallah, A.E., Abdallah, E.E., Obeidat, I.: An energy-efficient threshold-based clustering protocol for wireless sensor networks. Wirel. Pers. Commun. 1–14 (2013)
7. Aslam, N., Phillips, W., Robertson, W., Sivakumar, S.: A multi-criterion optimization technique for energy efficient cluster formation in wireless sensor networks. Inf. Fus. **12**(3), 202–212 (2011)
8. Badica, C., Scafes, M., Ilie, S., Badica, A., Muscar, A.: Dynamic negotiations in multi-agent systems. In: ICT in Education, Research and Industrial Applications: Integration, Harmonization and Knowledge Transfer, p. 8 (2011)
9. Shah, K., Kumar, M.: Distributed independent reinforcement learning (dirl) approach to resource management in wireless sensor networks. In: IEEE International Conference on Mobile Adhoc and Sensor Systems, 2007. MASS 2007, pp. 1–9. IEEE (2007)
10. Shah, K., Di Francesco, M., Kumar, M.: Distributed resource management in wireless sensor networks using reinforcement learning. Wirel. Netw. 1–20 (2012)

11. Shah, K., Di Francesco, M., Anastasi, G., Kumar, M.: A framework for resource-aware data accumulation in sparse wireless sensor networks. Comput. Commun. **34**(17), 2094–2103 (2011)
12. Barto, A.G.: Reinforcement learning: An introduction. MIT Press (1998)
13. Dimarogonas, D.V., Johansson, K.H.: Event-triggered control for multi-agent systems. In: Proceedings of the 48th IEEE Conference on Decision and Control, 2009 held jointly with the 2009 28th Chinese Control Conference. CDC/CCC 2009, pp. 7131–7136. IEEE (2009)
14. Devillé, M., Le Borgne, Y.A., Nowé, A., De Causmaecker, P., Maervoet, J., Messelis, T., Verbeeck, K., Vermeulen, T.: Reinforcement learning for energy efficient routing in wireless sensor networks. In: Proceedings of the 23rd Benelux Conference on Artificial Intelligence, pp. 89–96 (2011)
15. Forster, A., Murphy, A.L.: Clique: role-free clustering with q-learning for wireless sensor networks. In: 29th IEEE International Conference on Distributed Computing Systems, 2009. ICDCS'09, pp. 441–449. IEEE (2009)

Chapter 8
Technical Challenges and Implementation of Mobile Ad-hoc Water Level Measuring System

Takanobu Otsuka, Yoshitaka Torii and Takayuki Ito

Abstract In recent years, and landslides caused by flooding of the river due to heavy rain, the number of such flooding of houses has increased. With regard to internal water damage, the structure and of the river, above sea level and, such as where a landfill, how completely prevent flood damage by internal water is difficult. To that end we have done a prototype of a portable ad hoc simple water gauge for the purpose of installation in the field in collaboration with Nippon Koei Co., Ltd. In particular, in a portable of the wireless sensor networks, communication reliability is greatly influenced by the surrounding environmental factors by communication node movement. Furthermore, features and issues required portable ad hoc simple water gauge system is operated in the actual disaster site, it is necessary to determine standing worker eyes for performing production. In this paper, it is possible to organize the task of ad hoc simple water gauge system that assumes the movement, was carried out implementation of portable wireless sensor networks.

Keywords Wireless sensor network · Field informatics · Disaster informatics

8.1 Introduction

In recent years, and landslides caused by flooding of the river due to heavy rain, the number of such flooding of houses has increased. With regard to internal water damage, the structure and of the river, above sea level and, such as where a landfill, how completely prevent flood damage by internal water is difficult. At present, it is only possible addressing of Operating the drainage devices typified by car to pump submerged location. However, in order to quickly drained work as possible, it is necessary to understand the full flooding water by the water level data of the flooding point throughout. To that end we have done a prototype of a portable ad hoc simple water gauge for the purpose of installation in the field in collaboration with

T. Otsuka (✉) · Y. Torii · T. Ito
Nagoya Institute of Technology, Nagoya, Japan
e-mail: otsuka.takanobu@nitech.ac.jp
URL: http://www.itolab.nitech.ac.jp

Q. Bai et al. (eds.), *Multi-agent and Complex Systems*, Studies in Computational Intelligence 670, DOI 10.1007/978-981-10-2564-8_8

Nippon Koei Co., Ltd. In particular, in a portable of the wireless sensor networks, communication reliability is greatly influenced by the surrounding environmental factors by communication node movement. Therefore, differ from the fixed wireless sensor networks. It, the radio wave condition and buildings around which changes due to the movement, is the effect of such plants. Furthermore, features and issues required portable ad hoc simple water gauge system is operated in the actual disaster site, it is necessary to determine standing worker eyes for performing production. In this paper, it is possible to organize the task of ad hoc simple water gauge system that assumes the movement, was carried out implementation of portable wireless sensor networks. The abstract goes here.

The rest of the paper is organized as follows. Section 8.2 introduces previous studies and the relative position of our research. Section 8.3 describes the configuration of our system. Then, describe the suggested design guidelines for overcoming the problems in Sect. 8.4, in Sect. 8.5 indicate experimental results of the prototype system. And finally, Sect. 8.6 in show report summary and future challenges.

8.2 Related Work

8.2.1 Research Trends in WSN

Wireless sensor networks (WSN) in recent years, IoT (Internet of Things), M2M (Machine to Machine) is a technology that forms the core of, research has been conducted widely [1]. Nodes constituting the WSN, the temperature, illumination, acquires the sensor data such as acceleration, can constitute a "multi-hop ad hoc network" to transfer a bucket relay system the obtained data by using a radio wave [2, 3]. WSN, in order to be able configure the autonomous network simply by placing nodes, can reduce the installation work in the field use. Also, it is possible to obtain the sensor data, it is possible to capture real-world dynamics, tracking and monitoring of the natural environment of the object have been widely studied as a promising application in WSN.

Wireless sensor networks are a number of slave units and a sink node for aggregating information from the (sensing node) (Sink mode), and is composed of a multi-hop in the information of the relay capable router nodes (Router node), the observation environment information and, it is beginning to be widely used in the fields such as smart home. In research applications, is famous MOTE [4] that was developed at UC Berkeley in 1998. In recent years has been standardized as IEEE802.15.4, have been actively studied with a Zigbee network. However, sales that have been licensed in Japan, construction authentication both unnecessary communication module, the output as compared to other countries has been severely limited, a large limitation in communication distance.

In recent years, individuals interoperable environment information acquisition device are sold widely used for research applications and home. Wireless sensor system is a MEMSIC's Eko system [5] research was employed using agricultural [6] also have been performed, and the poor in scalability. Moreover, as a product of the consumer, there is a rice Davis's Vantage series [7], because it is only capable of operating in a dedicated application, it is not suitable for large-scale data collection. As an example of the environmental information settling by experimental WSN, it is possible to install a large number of observation nodes to an active volcano, research [8] are also performed on the data collection for the purpose of such predictions eruptions.
We have developed a sensor network device and the server application that can collect the high density of information on a large scale, is doing a collection of environmental data. In addition, in the wireless sensor network field, research [9] about the optimization of sensor arrangement for the purpose of reducing the power consumption and the construction of the sensor node applications there is a study of applying the agent technology [10]. Furthermore, the aim of manageability of the operational aspect, it is possible to be prepared the nodes of the program according to the purpose, the research on facilitating the construction of a sensor network according to applications [11], and depending on the circumstances there is autonomously study that constitute the network topology [12]. Thus, the study of WSN as a framework for performing a flexible measurement is active.

8.2.2 Studies on the Radio Wave Quality Stabilization of WSN

Furthermore, the aim of manageability of the operational aspect, it is possible to be prepared the nodes of the program according to the purpose, the research on facilitating the construction of a sensor network according to applications [11], and depending on the circumstances there is autonomously study that constitute the network topology [12]. However, the most sought performance in WSN is a network of stability. That is, the purpose of continuous data collection, and by optimizing the placement of fixed sensor networks, studies to keep the quality of the network constant is widely. For example, using a radio wave level (RSSI values) of the topology formed by the nodes, studies to optimize the node arrangement in a predefined spatial [13]. To capture the node as a cluster, research [14] to optimize the communication route. By optimizing the position of the relay station for multi-hop between nodes, and the like studies to reduce communication errors [15]. Also, by arranging a plurality of sink nodes, and the like studies to optimize the position of the sink node [16].
However, any research is also a study for the purpose of optimization for "closed environment", securing the communication quality of the mobile type WSN the ambient environment changes dynamically. Also, it is difficult to apply directly to the portable sensor networks that operate in the real field. In particular, the sensor network that is constructed in a wide range for the purpose of obtaining environmental information, for. The need to communicate distance, and agriculture, in river management

operations, in advance measures the radio wave state around the installation site, on the preliminary survey, technique for installation in consideration the placement of the measurement nodes are common. However, installation of the above for the portable sensor networks for the purpose of operating in the field of measuring the radio wave state in advance is not realistic. In recent years, and landslides caused by flooding of the river due to heavy rain, the number of such flooding of houses has increased. With regard to internal water damage, the structure and of the river, above sea level and, such as where a landfill, how completely prevent flood damage by internal water is difficult. At present, it is only possible addressing of Operating the drainage devices typified by car to pump submerged location. However, in order to quickly drained work as possible, it is necessary to understand the full flooding water by the water level data of the flooding point throughout. To that end we have done a prototype of a portable ad hoc simple water gauge for the purpose of installation in the field in collaboration with Nippon Koei Co., Ltd. In particular, in a portable of the wireless sensor networks, communication reliability is greatly influenced by the surrounding environmental factors by communication node movement. Therefore, differ from the fixed wireless sensor networks. It, the radio wave condition and buildings around which changes due to the movement, is the effect of such plants. Furthermore, features and issues required portable ad hoc simple water gauge system is operated in the actual disaster site, it is necessary to determine standing worker eyes for performing production.

In this paper, it is possible to organize the task of ad hoc simple water gauge system that assumes the movement, was carried out implementation of portable wireless sensor networks.

8.3 Development and Technical Challenges of Portable WSN

8.3.1 Required to Portable Ad-hoc Water Gauge Capability

In recent years, and landslides caused by flooding of the river due to heavy rain, the number of such flooding of houses has increased. With regard to internal water damage, the structure and of the river, above sea level and, such as where a landfill, how completely prevent flood damage by internal water is difficult. At present, it is only possible addressing of Operating the drainage devices typified by car to pump submerged location. However, in order to quickly drained work as possible, it is necessary to understand the full flooding water by the water level data of the flooding point throughout. A mobile ad-hoc level gauge is mounted to the sink node to the pump truck to perform the drainage of flooded areas, and to aggregate the water level information from the measurement nodes installed anywhere. By their information, the water level conditions in the flooded area was visualized, of and calculation time, the development for the purpose of most water pump truck to high point to shorten the

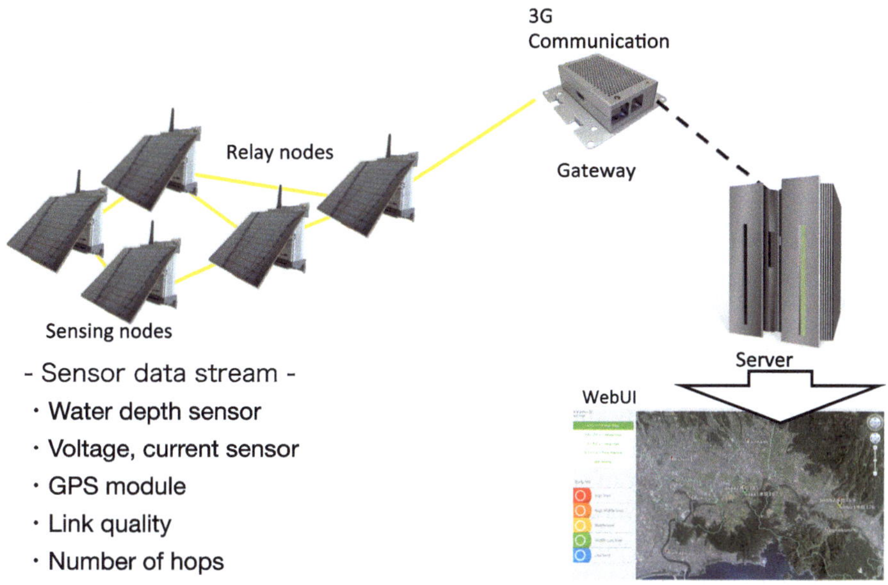

Fig. 8.1 System overview

time required for waste water by moving conducted required for drainage to have. In addition, the cooperation with the pump future vehicle integrated management system is intended to improve the efficiency of waste water work using a pump wheel with a small number of operators. We show system outline in Fig. 8.1. The system, for the system to be operational in reality, high reliability unlike sensor node of research applications are required. Further, because it differs with the agricultural WSN are WSN and outdoor operation fixed, the following three problems are present.

Challenge of communication reliability Since the sink node, the measurement nodes are installed anywhere, it is difficult to consider the pre-position. Therefore, not only the communication distance required there to check the redundancy of the ad hoc network.

Challenges of the systems Since the measurement node itself moves, the current position in real time and it is necessary to clearly display the water level change.

Challenges in operation Because it is operated at disaster sites, ease of installation in the field, and maintainability of the time of failure is essential.

Also, unlike the stationary WSN, likely to external factors affecting the signal quality is changed dynamically, as in the fixed WSN, the advance radio wave condition is measured, and to optimize the placement is difficult. Therefore, that all communication nodes act as a relay station by the ad hoc network is required. Thus, rather than the hardware side, and identifying the measurement nodes and the relay node by setting the connection whether the water level gauge on the server software. Moreover, since the measurement nodes, the task of mounting the router nodes located disaster site, it is necessary to reduce the effort of the operator. Therefore, of course ease of installation, visualization of communication nodes location and, by performing the water level information visualization, it is required it is easy to use system for the site worker.

8.4 Proposal and Implementation of Technical Problem Solving

In this chapter, challenges and solve problems listed in Sect. 8.3, and the implementation are described. First, we describe the communication reliability challenges. The communication reliability, this is a problem of how to overcome the communication failure due to the influence of the buildings and forest zone in the disaster site. When compared to the number of sensor network that is installed in a room environment, the outdoor sensor network to be installed in areas where buildings are crowded, large influence of radio wave shield by the position to be installed, if it is impossible to communicate is assumed. In particular, in our development to have portable of wireless sensor networks, communication reliability is greatly influenced by the surrounding environmental factors by communication node movement. Therefore, unlike the fixed wireless sensor network, the radio wave condition and buildings around which changes the movement, many of the differences that including the effects of such plants present. We, many of the fields making a wave test in at, in order to solve the signal quality issues are important in the portable sensor networks, it is also proposed recommendation algorithm installation position. Also, in the WSN research field in Japan using ad-hoc network, but Digi's Xbee series is often used, in association studies we conducted, and significant advantages in power consumption and communication distance is not seen. We taking advantage of the knowledge, as a communication module, as well as adopting the Tokyo Cosmos's TWE-Strong [17], can be switched the relay station and the measurement station software that you can operate all of the node as a relay node and the measurement node I was adopted, such architecture. Relay node or sensor node to be switched from webUI, with respect to the conversion of the depth data to the water level, and is displayed as the water level data by using the offset amount from the reference value. Therefore, even for the place radio wave condition is bad, it is possible to ensure the reliability of communication by relaying data by the relay node. Also, the problem of the system plane, with respect to locating the node was it possible to send the current position to a GPS module. The sensor management screen is shown Fig. 8.2. In addition, to take advantage of the knowledge [5, 18] of the system being operational in agricultural in the challenges of operational aspects, in the event of sensor failure by a modular structure for circuit, and module replacement in I was easily recoverable structure. Actually implemented was the structure of the circuit board I is shown in Fig. 8.3. The discussion of the operators with respect to the mounting of the communication node, and are easily removable facilitating structure that is fixed to the tripod or the like

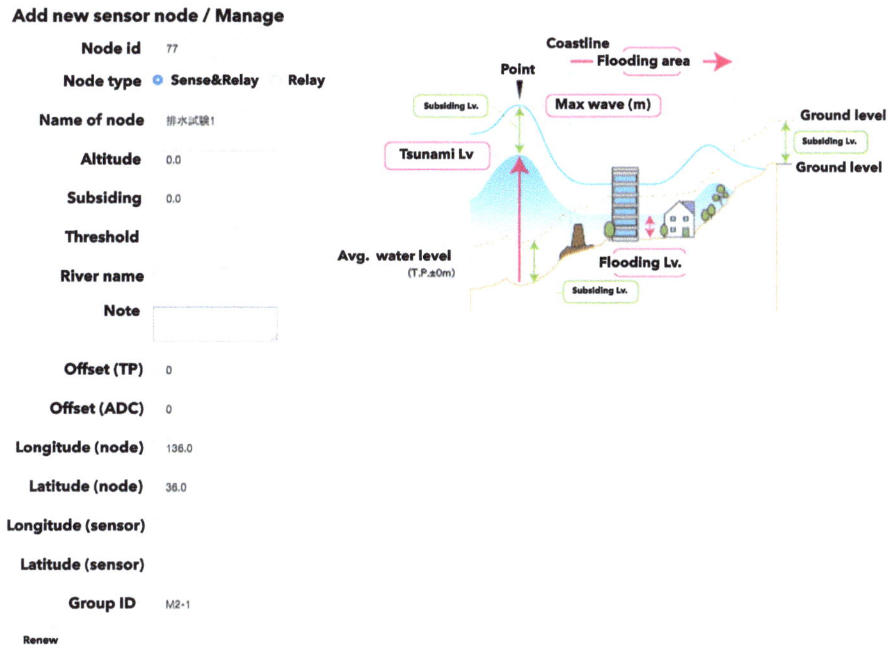

Fig. 8.2 Switching of measurement and relay node by webUI

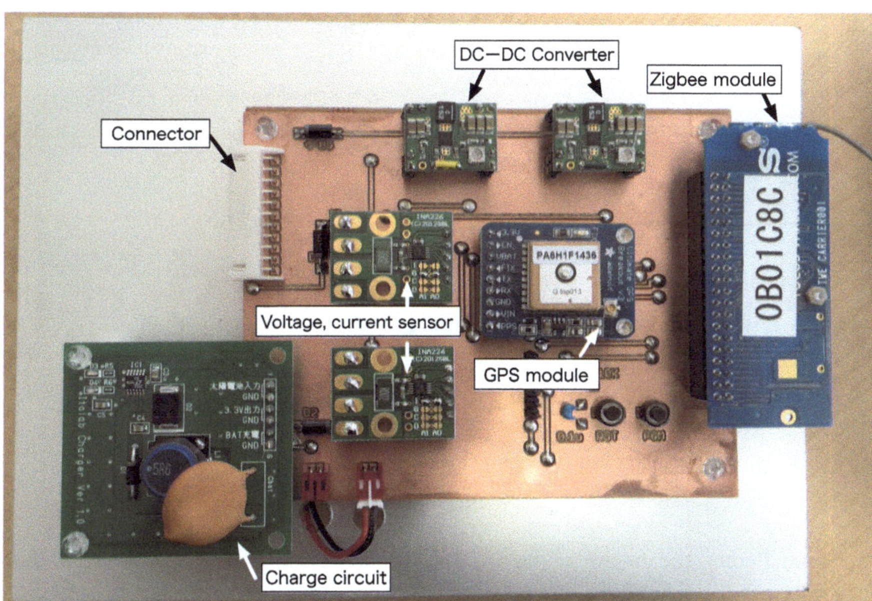

Fig. 8.3 The developed circuit board

Fig. 8.4 Appearance of the communication node

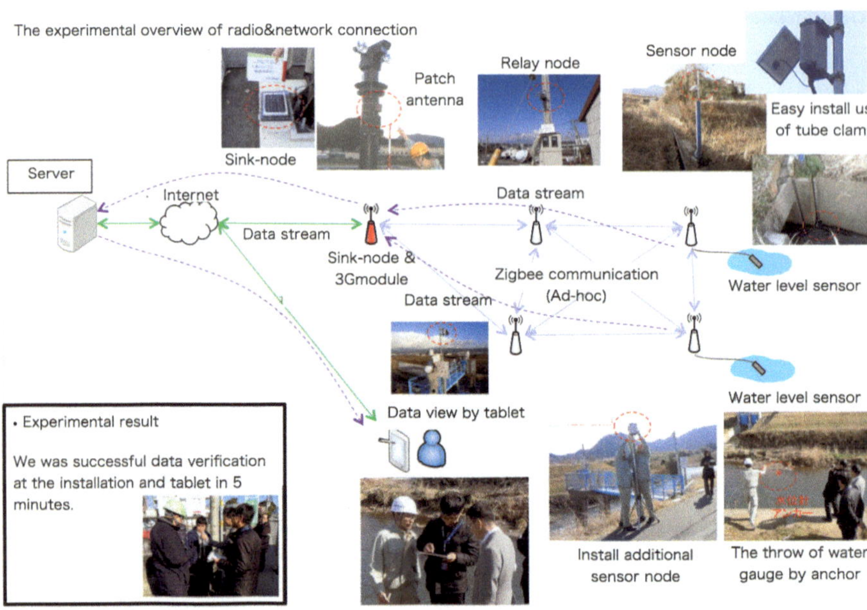

Fig. 8.5 Actual overview of test fields

that is installed on the ground using a clamping tube. Also, with respect to solar cells that are required in the case of long-term operation, by using a hinge structure for delivery easier, thereby achieving the efficiency of operation as required. We show the communication node appearance in Fig. 8.4. As described above, by performing the actual design and implementation standing operational point of view, a system easy to use for operators were implemented in hardware and software both sides.

8.5 The Operation Experimental Results with Actual Field

In this section, we show the results of operational experiments were performed at Mishima, Shizuoka on February 12, 2015. This experiment fields, are present in the branch line and the main line to the confluence position of Kano, it has experienced three times of flooding over the past 10 years, is designated as a flood hotspots in hazard map provided by Mishima that has been is a region. This experiment was carried out under the supervision of which is the main purchaser Ministry of Land, Infrastructure and Transport Chubu Maintenance Division. The equipment layout in the experimental field, and for the test item is shown in Fig. 8.5. Experiments, assuming the inland water damage occurs, we observed the actual level due to installation and immersion depth sensor type portable ad hoc water gauge using the actual river. In this experiment it was confirmed whether satisfies the following performance.

- After local arrival, immediately confirmed that it is possible to operate.
- Confirmation of redundancy due to the ad hoc network when a part of the communication path becomes disconnected.
- Easiness of tablet viewing by webUI.

In addition to check these items, experiments were carried out at actual river watershed. Portable ad hoc water gauge is mounted in the following procedure.

1. Fixed the communication node by tube clamp on a tripod.
2. Writes tow anchor rivers, to irrigation canals.
3. And turned on by the mounted pilot wire water depth sensor to the anchor.
4. Connect the water depth sensor to communication node.

the above procedure, it is possible to measure water level anywhere. Actual installation landscape and the steps Fig. 8.6. Established a gateway that is connected to the 3G communication network that assumes the pump car mounted on the site, and transmits the data to the web server. Transmitted data is processed on the server, the position of the measurement nodes by performing a visualization and are visualized water level information. Visualized data by the server software can be viewed by using a tablet, it is possible to check the status of the internal water damage to real time. Therefore, efficient operation of the pump car is made possible. The data display example obtained in actual operation experiment I is shown in Fig. 8.7. Sampling interval of measurement is once every 30 s. We have confirmed that it satisfies the following capability by the experimental results.

Fig. 8.6 Actual installation landscape and procedures

1. After unloading equipment, can be measured water level in 5 min.
2. Even when it is intentionally cut off a portion communication path 30 s (one sampling period) within it communicatively detour to other routes.
3. About webUI, it can be viewed real data without having to explain many of the participants.

As shown above, the result of the operation experiment of this system, from the Ministry of Land, Infrastructure and Transport is the ordering party, has been developed by the present study, a portable ad hoc water gauge for operational possible internal water damage grasped at a disaster site is useful were evaluated as is. In addition, through the work in the field, or a need to change the wing nut so that it can be

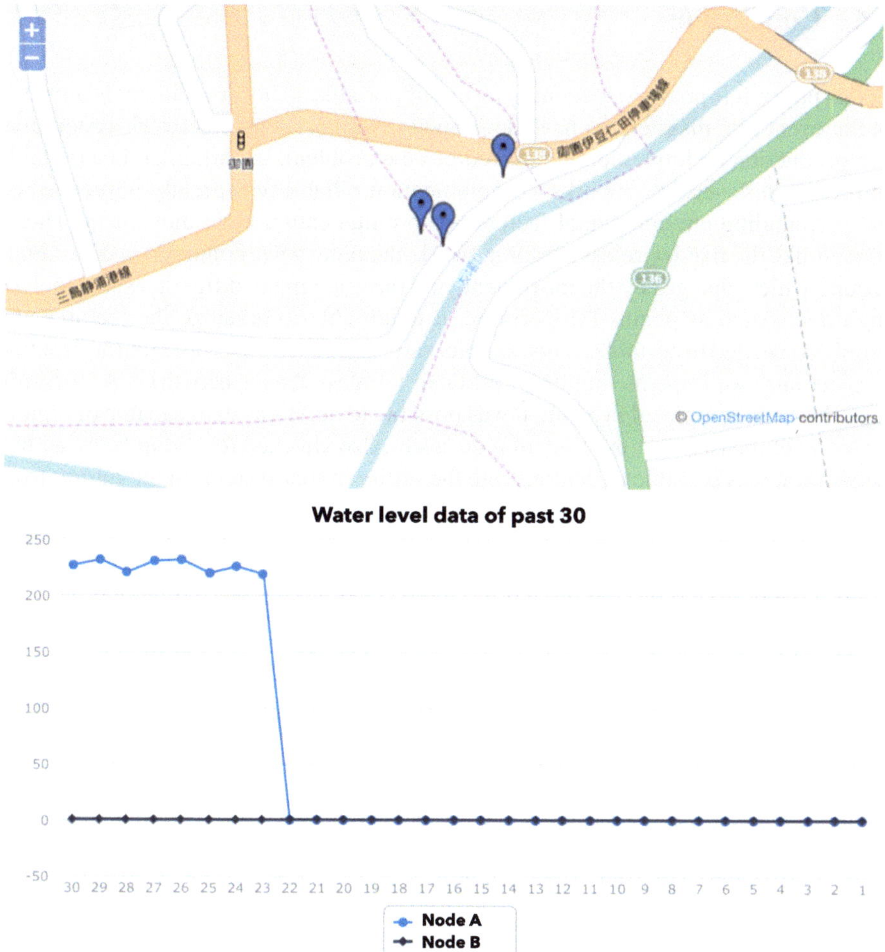

Fig. 8.7 The data display example by webUI

tightened by hand screwing part of the tripod, etc. becomes impossible communication in theory on the communication where possible, also a problem with the actual field operation can be extracted. In particular radio propagation in urban areas is often to determine the communication availability in concept sight, and the impact of the type waves reflected by the outer wall of the building, also been found that there are many parameters to determine the communication availability. Currently, we conduct ongoing operational test of the system, but some in the state has been fixed since March 2015, in Mishima, Shizuoka Misono district, using the node that was developed in this study, continuously water level I have continued to measure the data. By continuing test results, we have been continuously verify the reliability of long-term behavior of the hardware and software.

8.6 Conclusion

In this paper, it is possible to identify issues of portable ad hoc level meter is operated in the event of a disaster, we have developed a portable sensor network system that can production by implementing overcomes the problem. In particular, in a portable of the wireless sensor networks, communication reliability is greatly influenced by the surrounding environmental factors by communication node movement. Therefore, unlike the fixed wireless sensor network, the radio wave condition and buildings around which changes by the movement, and there are many differences that including effects such as plants. Furthermore, it is possible to organize the features and problems needed to portable ad hoc simple water gauge system is operated at an actual disaster site, we implementation of portable wireless sensor networks. As a result, installation is completed in 5 min, it was possible to build a system capable of rapidly water level measured at the time of a disaster. Also checked for redundancy ad hoc network, it was possible to demonstrate the utility of this system. In the future, from the fact that you're getting the water level data is currently ongoing in the real field, organize and challenges at the time of long-term investment, you are planning the size of the communication node. The next year, systems and cooperation with the actual pump trucks, and is supposed to perform the test in many real field, it will finish more useful system. The wireless communication standard is constantly advancing, has been also appeared higher performance communication module. Thus, changes or to a different communication module, so that aim to reduce power consumption due to software and hardware control, so that the sensor network is operated in a disaster can be easily realized, will be performed by continuing research and development.

Acknowledgments Part of this study, and the Ministry of Land, Infrastructure and Transport Chubu Regional Development Division is a part of the "Nobi Plain of drainage plan". In addition, it is due to the collaboration with Nippon Koei Co., Ltd.

References

1. Akyildiz, I.F., Su, W., Sankarasubramaniam, Y., Cayirci, E.: Wireless sensor networks: a survey. Comput. Netw. **38**, 393–422 (2002). Elsevier Science
2. Chong, Chee-Yee, Kumar, Srikanta P.: Sensor networks: evolution, opportunities, and challenges. Proc. IEEE **91**(8), 1247–1256 (2003)
3. Kuorilehto, M., Hannikainen, M., Hamalainen, T.D.: A survey of application distribution in wireless sensor networks. EURASIP J. Wirel. Commun. Netw. **5**(5), 774–788 (2005)
4. Wireless Sensor Network MOTE. http://www.xbow.jp/01products/index.html
5. MEMSIC .inc, Wireless Sensor Network, "Eko". http://www.memsic.com/wireless-sensor-networks/
6. Ito, M.: Environmental monitoring project with sensor network technology. IEEE Jpn. **2010**(66) (2010)
7. Davis Instruments, Whether Products. http://www.davisnet.com/weather/
8. Werner-Allen, G., Lorincz, K., Welsh, M., et.al.: Deploying a wireless sensor network on an active volcano. IEEE Internet Comput. **2006**, 18–25 (2006)

9. Lal, D., Manjeshwar, A., Herrmann, F., Uysal-Biyikoglu, E., Keshavarzian, A.: Measurement and characterization of link quality metrics in energy constrained wireless sensor networks. In: Proceedings of IEEE International Conference on Global Communications Conference, pp. 446–452 (2003)
10. Kenta, C., Akihiko, O., Shinichi, H.: Intelligent mobile agent framework for multi-purpose wireless sensor network applications. (Mobile Applications, Special Issue, Mobile Communications and Intelligent Transportation Systems towards the Ubiquitous Era) IPSJ J. Inf. Process. Soc. Jpn. (IPSJ)
11. Hill, J.,Szewczyk, R., Woo, A., Hoilar, S., Culler, D.E., Kristofer, S.,Pister, J.: System architecture directions fbr networked sensors architectural support for programming languages and operating system **3**, 93–104 (2000)
12. Fbk, C.L., Roman, G.-C., Lu, C.: Rapid development and flexible deployment of adaptive wireless sensor network applications. In: Proceedings of 25th IEEE International Conference on Distributed Computing Systems Vbl, vol. 12, pp. 653-662 (ICDes, 05) (2000)
13. Romer, Kay, Mattern, Friedemann: The design space of wireless sensor networks. IEEE Wirel. Commun. **11**(6), 54–61 (2004)
14. Younis, Ossama, Krunz, Marwan, Ramasubramanian, Srinivasan: Node clustering in wireless sensor networks: recent developments and deployment challenges. IEEE Netw. **20**(3), 20–25 (2006)
15. Hongyang, C.H.E.N., et al.: An improved DV-Hop localization algorithm with reduced node location error for wireless sensor networks. IEICE Trans. Fund. Electron. Commun. Comput. Sci. **91**(8), 2232–2236 (2008)
16. Akihide, U., Hisao, Y.: Particle swarm optimization for computing plural acceptable solutions and its application to sink node allocation problem in wireless sensor networks with multiple sinks. In: The IEICE Transactions on Information and Systems (Japanese edition). The Institute of Electronics, Information and Communication Engineers
17. Tokyo Cosmos CO. LTD, TWE-Strong. http://tocos-wireless.com/jp/products/TWE-001STRONG.html
18. Masayuki, H.: Global sensor network with field servers technical report of IEICE. The Institute of Electronics, Information and Communication Engineers

Part II
Intelligent Agents and Their Applications

Chapter 9
Membership Function Based Matching Approach of Buyers and Sellers Through a Broker in Open E-Marketplace

Dien Tuan Le, Minjie Zhang and Fenghui Ren

Abstract A broker in a market enables buyers and sellers to do business with each other and can provide many value-adding functions that cannot be replaced by direct buyer-seller dealings. Recently, some research has focused on this issue. However, broker modelling based on buyer's membership functions to carry out a matching process between buyer's requirements in fuzzy preference information and seller's offers is still sparse. Thus, this paper proposes membership function based matching approach of buyers and sellers through a broker in open e-marketplace. The major contributions of this paper are that (i) a proposed framework is applicable to help a broker to carry out the matching process between buyers and sellers; (ii) a proposed method is to determine buyer's attribute weight with soft constraints by using association rule mining; and (iii) an objective optimization function and a set of constraints are built to help a broker to maximize buyer's total utility. Experimental results demonstrate the good performance of the proposed approach in terms of satisfying buyer's requirements and maximizing buyer's total utility.

Keywords Matching approach · Seller's offers · Buyer's requirements · Buyer's total utility

9.1 Introduction

Research on brokers or intermediaries in the markets as the third party of the trading processes in e-markets has been a very active direction in recent years. Li et al. [9] developed a mathematical model to solve the multi-attribute matching problem

D.T. Le (✉) · M. Zhang · F. Ren
School of Computing and Information Technology, University of Wollongong, Wollongong, NSW 2500, Australia
e-mail: dtl844@uowmail.edu.au

M. Zhang
e-mail: minjie@uow.edu.au

F. Ren
e-mail: fren@uow.edu.au

Q. Bai et al. (eds.), *Multi-agent and Complex Systems*, Studies in Computational Intelligence 670, DOI 10.1007/978-981-10-2564-8_9

through a matchmaker. Jiang et al. [7] proposed a novel matching approach for a broker to achieve the optimal trade matching in multi-attribute exchanges under consideration of the trading volume and the matching degree. Alpar [1] developed a conceptual framework of matching in B2B e-marketplaces environments and proposed the new algorithm for the implementation of the functionalities of the matchmaker. Blume et al. [2] studied the trading processes in general e-markets between buyers and sellers through a layer of intermediaries. Jung et al. [8] proposed a two-layered multi-agent framework to match between buyers and sellers through brokerage by using constraint satisfaction problems (CSP).

Although the above approaches have focused on studying brokers as a third party in the trading processes between buyers and sellers, there is little theory and few guidelines to help a broker to optimize the trading matching between buyer's requirements in fuzzy preference information and seller's offers. Therefore, following challenges for broker modelling still exist, which are (i) how to map buyer's requirements to seller's offers optimally; (ii) how to maximize buyer's total utility under consideration of buyer's requirements in fuzzy preference information and seller's offers; and (iii) how to determine buyer's attribute weight with soft constraints based on historical trading dataset to support broker's decision.

In order to solve the above challenges, this paper proposes membership function-based matching approach in multi-attribute exchanges through a broker between buyers and seller. The major contributions of this paper are as follows. (i) The design of membership function based matching approach in multi-attribute exchanges is in general level by considering general markets so that it can be applied to different types of markets; (ii) The proposed method is to derive buyer's attribute weight with soft constraints by using association rule mining; and (iii) An objective optimization function and a set of constraints are proposed to maximize buyer's total utility in regard to buyer's requirements and seller's offers. Experimental results demonstrate the good performance of the proposed approach in terms of satisfying buyer's requirements and maximizing buyer's total utility.

The rest of this paper is organized as follows. Problem description is presented in Sect. 9.2. The proposed matching approach in the markets is introduced in Sect. 9.3. An experiment is presented in Sect. 9.4. Section 9.5 compares our approach with some related work. Section 9.6 concludes in this paper and points out our future work.

9.2 Problem Description

There are three members in the trading process with multi-attribute exchanges, i.e., buyers, sellers and a broker. The trading process is shown in Fig. 9.1. The broker is often called the facilitator, who acts as an intermediary between the buyer and the seller in the commodity exchange. In this paper, the broker's responsibility is to match n ($n \geq 1$) buyers with m ($m \geq 1$) sellers for the same commodity with multi attribute exchanges in order to satisfy buyer's requirements. Buyer b_i ($i = 1, 2, \ldots n$)

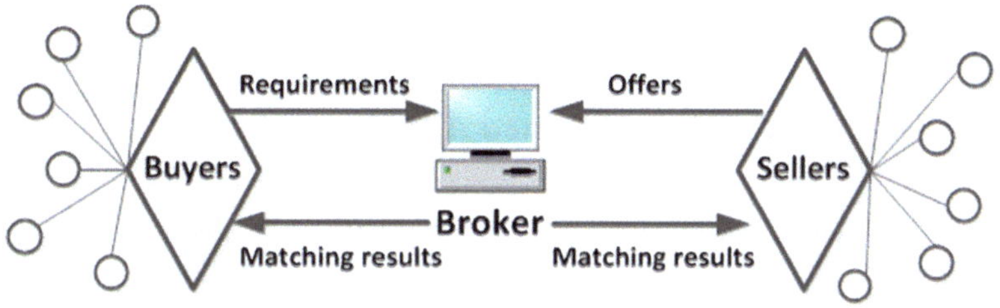

Fig. 9.1 The trading processes through a broker in open E-marketplace

and seller s_j $(j = 1, 2, \ldots m)$ have a single unit of the commodity with multiple attributes to buy or sell. Multi attributes in buyer's requirements are divided into two categories including the attributes with hard constraints and the attributes with soft constraints. The attributes with hard constraints mean that their constraints are presented in the form of an 'equal to' notation while the attributes with soft constraints are presented in the form of inequality and their constraints can be relaxed within the given scope of values [8].

From the buyer's part, buyer b_i can present b_i's requirements through many attributes. In general, when buyers select a certain product from the markets through a broker, buyers work with product's uncertain information or product's attribute level choices. Under these situations, it is difficult for buyers to estimate the attribute levels with exact numerical values. Thus, buyers normally express their requirements of the product features in fuzzy or linguistic terms [3]. For example in a washing machine purchasing problem, buyer's preference information related to price, popularity, comfort and maintenance cost is sent to a broker in following terms.

Price: The price of washing machine should be *around AUD1,000.*

Popularity: Popularity of washing machine *should be high.*

Comfort: Overall washing machine *should be comfortable.*

Maintenance: Maintenance cost of washing machine *should be medium.*

Fuzzy or linguistic terms are the italic words in the above example. The price attribute can be presented through fuzzy numbers while the other attributes, i.e., popularity, comfort and maintenance cost can be expressed by using the fuzzy or linguistic terms [6].

Similarly, from seller's point of view, seller s_j's offer is related to many attributes. Level of each attribute in s_j's offer is provided in details to a broker.

Based on the above analysis, a key problem is how to help a broker to find the optimal matching pairs so that buyer's requirements are satisfied and buyer's total utility is maximized. Therefore, the proposed matching approach is to solve this problem and presented in Sect. 9.3.

9.3 The Proposed Matching Approach

9.3.1 Framework of the Proposed Approach

The framework of the proposed approach presented in Fig. 9.2. helps a broker to solve the matching problem between buyer's requirements in fuzzy preference information and seller's offers with multi-attribute exchanges. The proposed approach focuses on maximizing buyer's total utility through a broker under business environments.

In the framework, buyer's requirements in fuzzy preference information and seller's offers related to a multi-attribute commodity are submitted to a broker. A broker communicates with buyers by using the direct rating (point estimation) method to build buyer's membership function for each attribute. Based on buyer's membership function, a broker calculates buyer's utility for each attribute as per seller's offers to determine a constraint satisfaction layer. The constraint satisfaction layer includes sellers which satisfy at least a certain buyer's requirements. Then, a broker uses association rule mining to estimate buyer's attribute weight with soft constraints based on buyer's historical trading datasets. After that, an objective optimization function and a set of constraints are generated to maximize buyer's total utility. Finally, the objective optimization function is solved by linear programming problem (LPP) to obtain the optimal matching pairs. The main issues of the proposed approach, i.e., building the calculation of buyer's utility, calculating buyer's attribute weight with soft constraints and building the objective optimization function are presented in details in the following subsections.

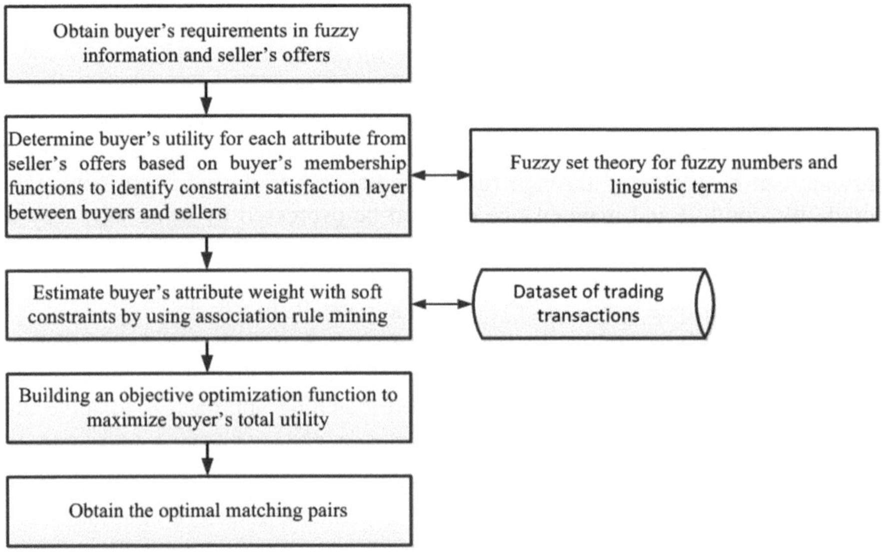

Fig. 9.2 The framework of the broker modeling approach

9.3.2 Building the Calculation of Buyer's Utility

In the majority of market settings and products, buyer's own preferences about products and their features are normally expressed in a qualitative or linguistic manner because buyer's knowledge about products is relatively vague. Thus, it is difficult for buyers to express their preferences with an exact numerical value. On the other hand, the use of words or sentences rather than numbers enables a more flexible and realistic form of adequately expressing day-to-day business terms. To estimate buyer's preferences in a qualitative or linguistic manner, fuzzy set theory is best suited to deal with the qualitatively defined terms (linguistic assessments) in a quantitative manner [12]. A precise definition of fuzzy set is as follows:

Definition Let X be a set of objectives. A fuzzy set A in X is defined as a set of ordered pairs $A = \{x, \mu_A(x)\}$, where $\mu_A(x)$ represents a membership function of fuzzy set A, which associates each point $x \in X$ with a real number in the interval [0,1]. The value $\mu_A(x)$ is called the grade of membership of x in A.

Let a set of buyer $\mathbf{B} = \{b_1, b_2, \ldots, b_n\}$ and a set of sellers $\mathbf{S} = \{s_1, s_2, \ldots, s_m\}$, buyer's requirements and seller's offers are related to many attributes which can be split into a set of the attributes with soft constraints $\mathbf{A} = \{a_1, a_2, \ldots, a_k\}$ and a set of the attributes with hard constraints $\mathbf{H} = \{h_1, h_2, \ldots, h_z\}$ [7]. Let a set of constraint values $\mathbf{c_i} = \{c_{i1}, c_{i2}, \ldots, c_{iz}\}$ and $\mathbf{c_j} = \{c_{j1}, c_{j2}, \ldots, c_{jz}\}$ for the attributes with hard constraints in b_i's requirements and s_j's offers, respectively. Similarly, a_{il} and a_{jl} denote the attribute level l with soft constraints of buyer b_i and seller s_j, respectively. Buyer's requirements related to the attributes with hard constraints must be satisfied by seller's offers to attend broker's matching process. Furthermore, the nature of fuzzy information is related to the attributes with soft constraints and is not allowed for hard attributes. Thus, the procedure of calculating buyer's utility for the attributes with soft constraints is presented as follows:

Step 1: A broker receives b_i's product requirements in fuzzy preference information and s_j's offers in terms of its attributes. A broker determines buyer's membership function for each attribute by using the direct rating (point estimation) method [10]. In this method, a broker communicates with buyers to determine buyer's preference point through questions. Broker's questions require a buyer to select one point on the reference axis (using numerical or verbal scale) that best describes this element. For example, a broker starts the simplified interactive procedure with buyers to build a membership function for the attribute of price. It consists of 3 questions that allows to identify three reference points within the feasible range of price as follows:

- Question 1: "What is the worst option for the attribute of price?"→ "everything is the worst if price of product is more than or equal to 25 AUD".
- Question 2: "What is the perfect option for the attribute of price that would give you full satisfaction level?" → "the perfect price is less than or equal to 15 AUD".
- Question 3: "What is a medium resolution level for you with regard to price? → "an average 20 AUD".

Based on buyer's responses above, the continuous membership function of the price attribute is presented as follows:

$$\mu(x) = \begin{cases} 1 & for\ \ x \leq 15 \\ \frac{25-x}{10} & for\ \ x \in \langle 15, 25\rangle \\ 0 & for\ \ x \geq 25 \end{cases}$$

In general, b_i's membership function for each attribute is presented as follows.

$$\{a_1^i, \mu_{a_1^i}\}, \{a_2^i, \mu_{a_2^i}\}, \ldots \{a_k^i, \mu_{a_k^i}\}, \tag{9.1}$$

where a_k^i represents the fuzzy set of the kth attribute for buyer b_i and $\mu_{a_k^i}$ presents the membership function of the fuzzy set corresponding to attribute a_k for buyer b_i.

Step 2: A broker determines buyer's utility for each attribute based on b_i's membership function and b_i's requirements. It is presented as follows:

$$\{(a_1^i, \mu^i(a_{i1})), (a_2^i, \mu^i(a_{i2})), \ldots (a_k^i, \mu^i(a_{ik}))\}, \tag{9.2}$$

where $\mu^i(a_{ik})$ is b_i's utility for attribute a_k, $\mu^i(a_{ik})$ is determined from b_i's requirements and b_i's membership function so it means that b_i expects to find out the minimal utility value $\mu^i(a_{ik})$.

Step 3: Based on b_i's membership function for each attribute, a broker determines b_i's utility for each attribute as per s_j's offer. It is presented as follows:

$$\{(a_1^i, \mu^{ij}(a_{j1})), (a_2^i, \mu^{ij}(a_{j2})), \ldots (a_k^i, \mu^{ij}(a_{jk}))\}, \tag{9.3}$$

where $\mu^{ij}(a_{jk})$ is b_i's utility for attribute a_k if s_j's offer is provided to b_i.

Step 4: A broker determines a constraint satisfaction layer by comparing $\mu^{ij}(a_{jl})$ with $\mu^i(a_{il})$, and c_{ig} with c_{jg}. More specifically, if $\mu^{ij}(a_{jl}) \geq \mu^i(a_{il})$ $(l = 1, 2, \ldots, k)$ and $c_{ig} = c_{jg}$(g=1,2,...,z) then seller s_j's offer satisfies b_i's requirements. Otherwise, seller s_j can not match with b_i.

9.3.3 Determining Buyer's Soft Attribute Weight

When carrying out the trading process between buyers and sellers in open environments, a broker needs to understand buyer's behavior in term of their attribute weight with soft constraints. Such understanding buyer's attribute weight with soft constraints helps a broker to retrieve buyer's real preferences. This will enable the broker to better understand buyers to select seller's appropriate offers to satisfy buyer's requirements. It is not an easy job to uncover buyer's attribute weight with soft constraints from fuzzy information. Our paper follows the Analytical Hierarchy

Process [11] to derive the attribute weight with soft constraints using association rule mining.

Assume that there are the number t of transactions ($\{T_1, T_2, \ldots, T_t\}$) carried out by buyer b_i so far. Each transaction consists of a set of sellers who provided a product to b_i. A broker determines b_i's soft attribute weight based on historical trading datasets as follows:

Step 1: For each transaction, a broker can find b_i's average utility for each attribute. For example, take the transaction T_s ($s \in t$) and assume that T_s includes s_1, s_2, s_3 and s_4. b_i's average utility T^i_{sl} of the attribute l in the transaction s is calculated as follows:

$$T^i_{sl} = (\mu^{i1}(a_{1l}) + \mu^{i2}(a_{2l}) + \mu^{i3}(a_{3l}) + \mu^{i4}(a_{4l}))/4 \tag{9.4}$$

Step 2: A broker calculates b_i's average utility of the lth attribute T^i_l in the entire business transactions as follows.

$$T^i_l = \frac{\sum\limits_{s=1}^{t} T^i_{sl}}{t} \tag{9.5}$$

Step 3: b_i's average utility T^i_{sl} of the attribute l in transaction s is checked to ensure that its value is at least equal to the value of T^i_l. If $T^i_{sl} < T^i_l$ then zero value is assigned to T^i_{sl}. Otherwise, T^i_{sl} is taken as T^i_{slnew}. This is necessary, as a broker does not want to consider b_i's average utility T^i_{sl} in any transaction if its value is less than b_i's average attribute utility of the entire business transaction.

Step 4: A broker calculates transaction frequency of the lth attribute T^i_{lnew} as follows.

$$T^i_{lnew} = \sum_{s=1}^{t} T^i_{slnew} \tag{9.6}$$

Step 5: Using the association rule mining [5], a broker can find the degree of association of the attribute a_l with any other attribute(s) w, where $w \in P(\mathbf{A} - a_l)$, $w \neq \varnothing$ and $P(\mathbf{A} - a_l)$ is a power set of any subset of $(\mathbf{A} - a_l)$. In particular, it is calculated as follows.

$$c^i_{lw} = \frac{\sum\limits_{s=1}^{t} [T^i_{slnew} \bigwedge T^i_{swnew}]}{T^i_{lnew}}, \tag{9.7}$$

where c^i_{lw} represents the degree to which b_i likes the attributes w because of the presence of the lth attribute ($l = 1, 2, \ldots, k$).

Step 6: A broker can calculate the degree of confidence of b_i for attribute a_l as follows:

$$c^i_l = \sum_{w \in P(\mathbf{A} - a_l), w \neq \varnothing} c^i_{lw} \tag{9.8}$$

Note that the number of non empty sets in $P(\mathbf{A} - a_l)$ is $2^{k-1} - 1$.

Step 7: If $rp^i_{ll'}$ $(l, l' = 1, 2, \ldots, k)$ represents b_i's relative preference of a_l over a'_l $(rp^i_{ll'} = c^i_l / c^i_{l'})$, then the matrix is generated as follows:

$$Z^i_{k,k} = \begin{bmatrix} rp^i_{1,1} & rp^i_{1,2} & \cdots & rp^i_{1,k} \\ rp^i_{2,1} & rp^i_{2,2} & \cdots & rp^i_{2,k} \\ \vdots & \vdots & \ddots & \vdots \\ rp^i_{k,1} & rp^i_{k,2} & \cdots & rp^i_{k,k} \end{bmatrix} \tag{9.9}$$

The eigenvector calculated from the maximum eigenvalue of the matrix $Z^i_{k,k}$ in Eq. (9.9) gives a broker buyer's attribute weight with soft constraints after the eigenvector is normalized. In particular, $\sum_{l=1}^{k} w^i_l = 1$, $w^i_l \geq 0$, where w^i_l is the weight of attribute a_l in b_i's requirements.

After determining buyer's attribute weight with soft constraints, a broker will build an objective optimization function to maximize buyer's total utility. The objective optimization function is presented in Sect. 9.3.4.

9.3.4 Building an Objective Optimization Function

Broker's decision making in open environments to maximize buyer's total utility through matching process between buyers and sellers is driven by the objective optimization function. Based on the problem description and notations, the objective optimization function and a set of constraints are built as follows:

$$f = \sum_{i=1}^{n} \sum_{j=1}^{m} \Big(\sum_{l=1}^{k} w^i_l \mu^{ij}(a_{jl}) x_{ij} \Big) \tag{9.10}$$

$$s.t. \sum_{i=1}^{n} x_{ij} \leq 1, j = 1, 2, \ldots, m \tag{9.11}$$

$$\sum_{j=1}^{m} x_{ij} \leq 1, i = 1, 2, \ldots, n \tag{9.12}$$

$$x_{ij} = 1, 0, (i = 1, 2, \ldots, n; j = 1, 2, \ldots, m) \tag{9.13}$$

$$\sum_{l=1}^{k} w^i_l = 1, (i = 1, 2, \ldots, n; l = 1, 2, \ldots, k) \tag{9.14}$$

$$x_{ij} = 0 \; if \; \mu^{ij}(a_{jl}) < \mu^i(a_{il}) (l = 1, 2, \ldots, k) \; or \; c_{ig} \neq c_{jg} (g = 1, 2, \ldots, z) \tag{9.15}$$

where the objective optimization function in Eq. (9.10) seeks to maximize the weight sum of buyer's utility, constraints (9.11) and (9.12) are that each buyer (seller) can buy (sell) one unit of the commodity at most. Constraint (9.13) is assignment variable constraints, if b_i matches with s_j, then $x_{ij} = 1$; otherwise $x_{ij} = 0$. Constraint (9.14) indicates b_i's attribute weight with soft constraints; and constraint (9.15) indicates a constraint satisfaction layer to attend broker's matching processes. Furthermore, Eq. (9.10) can be efficiently solved by well-known linear programming methods such as simplex methods or interior point method [4].

9.4 Experiments

In this section, we present our experimental results and analyse our matching approach's performance. The experiments mainly focus on testing maximizing buyer's total utility through matching between buyer's requirements and seller's offers. The rest of this section is divided into two subsections. Section 9.4.1 describes the experimental setting that have been applied in the experiments. Section 9.4.2 shows the experimental results and performance analysis in three different experimental scenarios.

9.4.1 Experimental Setting

In the experiments, we generate an artificial data of 10 buyers related to jacket's demand. Each buyer contains seven attributes: brand, price, delivery time, warranty time, size, colour and gender. From the buyer's point of view, brand, size, colour and gender are regarded as the attributes with hard constraints while price, delivery time and warranty time are considered as the attributes with soft constraints. Furthermore, each buyer includes 10 transactions selected from the historical trading dataset. Based on each buyer's historical trading dataset, a broker uses the association rule mining presented in Sect. 9.3.3 to determine buyer's attribute weight with soft constraints including price, delivery time and warranty time. Similarly, each seller contains seven attributes including brand, price, delivery time, warranty time, size, colour and gender. In the experiments, the proposed approach is evaluated under seller's market so the three different scenarios includes a number of different selected sellers. More specifically, a broker's matching approach is tested in three different scenarios presented in Table 9.1 to maximize buyer's total utility under different sellers.

Before the matching process is happened, a broker interacts with each buyer to determine buyer's membership function for each attribute with soft constraint by using the direct rating method presented in Sect. 9.3.2. Based on the buyer's responses, a broker is able to identify buyer's membership function for each attribute with soft constraint to carry out broker's matching process.

Table 9.1 Experimental scenarios

Scenario	Test purpose
1	To maximize buyer's total utility with 10 buyers and 5 sellers
2	To maximize buyer's total utility with 10 buyers and 10 sellers
3	To maximize buyer's total utility with 10 buyers and 20 sellers

9.4.2 Experimental Results and Analysis

In scenario 1, a broker uses the proposed matching to maximize buyer's total utility through finding out the allocations between buyers and sellers under considering that the number of buyers (10 buyers) is more than the number of sellers (5 sellers) in the markets. In general principle of markets, when buyer's demand is more than seller's supply, all buyer's requirements cannot be satisfied and it is difficult for buyers to obtain their high utility because a broker has a fewer opportunity to select seller's offers to satisfy buyer's requirements. The results of buyer's utility in scenario 1 are presented in Fig. 9.3 and the matching results are also presented in Table 9.2.

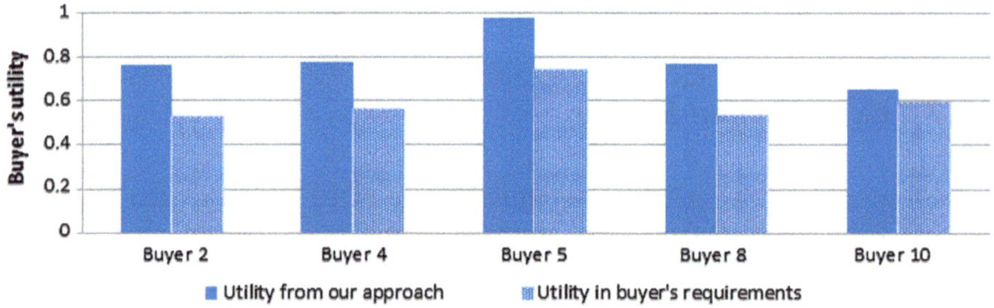

Fig. 9.3 Buyer's utility in Scenario 1

Table 9.2 Optimal matching pairs with the three different scenarios

	Scenario 1	Scenario 2	Scenario 3
1	$B_2 \longleftrightarrow S_5$	$B_1 \longleftrightarrow S_1$	$B_1 \longleftrightarrow S_7$
2	$B_4 \longleftrightarrow S_3$	$B_2 \longleftrightarrow S_5$	$B_2 \longleftrightarrow S_{20}$
3	$B_5 \longleftrightarrow S_2$	$B_3 \longleftrightarrow S_4$	$B_3 \longleftrightarrow S_2$
4	$B_8 \longleftrightarrow S_1$	$B_4 \longleftrightarrow S_3$	$B_4 \longleftrightarrow S_{10}$
5	$B_{10} \longleftrightarrow S_4$	$B_5 \longleftrightarrow S_8$	$B_5 \longleftrightarrow S_4$
6		$B_6 \longleftrightarrow S_2$	$B_6 \longleftrightarrow S_{12}$
7		$B_7 \longleftrightarrow S_{10}$	$B_7 \longleftrightarrow S_{17}$
8		$B_8 \longleftrightarrow S_7$	$B_8 \longleftrightarrow S_{15}$
9		$B_9 \longleftrightarrow S_9$	$B_9 \longleftrightarrow S_{19}$
10		$B_{10} \longleftrightarrow S_6$	$B_{10} \longleftrightarrow S_{14}$
	$f = 0.78$	$f = 0.82$	$f = 0.90$

Based on Fig. 9.3 and Table 9.2, it is clear that there are only five satisfied buyers including B_2, B_4, B_5, B_8 and B_{10} while other buyers do not satisfy. Our proposed approach through a broker helps five satisfied buyers to find their utility which is higher than utility in their requirements. However, five satisfied buyer's normalized total utility in scenario 1 is not high (0.78) because a number of sellers is less than a number of buyers.

Similarly, in scenario 2, a broker considers that the number of sellers is as equal as the number of buyers. Based on Fig. 9.4 and Table 9.2, it can be seen that buyer's requirements are also satisfied and the matching results are also found for each buyer. More specifically, buyer's normalized total utility in scenario 2 is relative high (0.82) and is higher than buyer's normalized total utility in scenario 1 because a broker has many opportunities to select seller's offers which satisfy buyer's requirements and increase buyer's total utility.

Finally, the number of sellers is twice as equal as the number of buyers. Based on Fig. 9.5 and Table 9.2, it is clear that except buyer's satisfied requirements, buyer's normalized total utility is very high (0.90) and higher than buyer's normalized total utility (0.78) in scenario 1 and buyer's normalized total utility (0.82) in scenario 2 because a broker in scenario 3 is more opportunity to select seller's offers which satisfy buyer's requirements than in scenario 1 and 2.

In summary, the proposed approach is perfectly performed under different situations in business environments. In general, if seller's supply is more than buyer's demand, a broker has many opportunities to choose seller's offers to satisfy buyer's requirements and increase each buyer's utility as well as buyer's total utility.

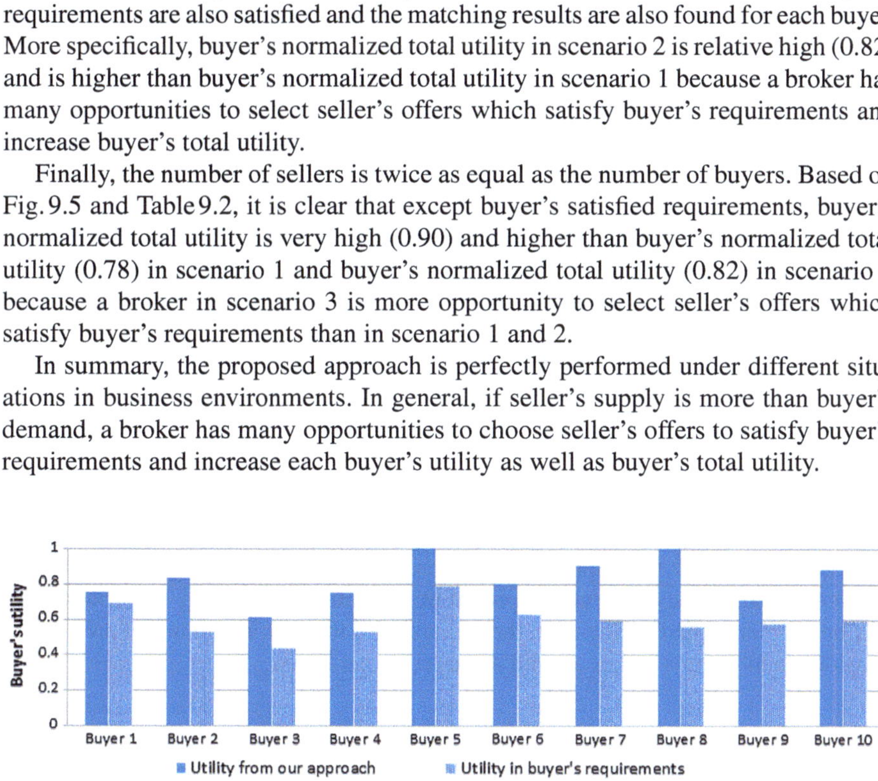

Fig. 9.4 Buyer's utility in Scenario 2

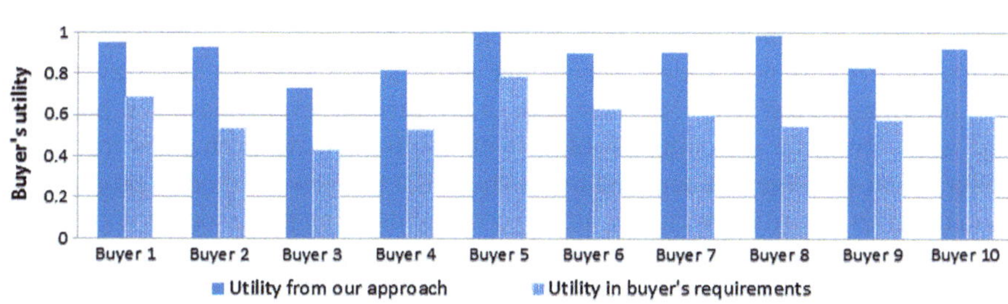

Fig. 9.5 Buyer's utility in Scenario 3

9.5 Related Work

There has been a lot of previous work on regarding the indirect interaction between buyer agents and seller agents through broker agents in e-markets. Jiang et al. [7] proposed a matching approach based on a bi-objective function to optimize the trade matching in multi-attribute exchanges with incomplete weight information through electronic brokerages (E-brokerages). In particular, the bi-objective optimization function is to maximize the matching degree and trading volume. The difference between Jiang's work and our work is that a broker in our approach uses the direct rating (point estimation) method [10] to communicate with buyers to determine buyer's membership function before a broker carries out the matching process between buyers and sellers. Thus, our approach is to maximize buyer's total utility through a broker based on buyer's membership function while Jiang et al. [7] does not pay attention to buyer's utility from its membership function in Jiang's bi-objective optimization function.

Li et al. [9] proposed two objective optimization functions to match buyers and sellers in B2B e-marketplace. The first and second objective optimization function are to maximize the total satisfaction of buyer and seller, respectively. Although buyer's attribute weight is considered in Li's multi objective function, buyer's attribute weight values are chosen by buyers. The novelty of our approach is that a broker determines buyer's attribute weight with soft constraints by using association rule mining based on historical trading datasets.

Jung et al. [8] modelled the trading phenomenon in the markets in which brokerage acted as a middleman between buyers and sellers. They proposed a two-layered mult-agent framework for brokerage between buyers and sellers. Based on buyer and seller's requirements, their approach helps brokerage to find out an optimal matching solution to satisfy buyer's various preferential requirements using constraint satisfaction problems (CSP). However, the limitation of their approach is that they do not consider each buyer's utility as well as buyer's total utility.

9.6 Conclusion and Future Work

This paper proposes the optimal matching method based on buyer's membership function through a broker in open e-marketplace. The proposed approach is novel because (1) it is a novel idea to consider buyer's requirements in fuzzy preference information and seller's offers with multi-attribute exchanges. The proposed approach solves the matching problem with multi-attribute exchanges through a broker based on buyer's membership function; (2) the new method is proposed to estimate buyer's attribute weight with soft constraints using the association rule mining; and (3) the objective optimization function and a set of constraints are generated to maximize buyer's total utility. The experimental results demonstrate the good performance for the proposed approach in aspects of satisfying buyer's requirements and buyer's total utility.

Future research includes extending the proposed approach to solve competition environments between brokers and dynamic environments.

References

1. Alpar, F.Z.: Matchmaking framework for b2b e-marketplaces. Inf. Econ. **14**(4), 164–170 (2010)
2. Blume, L., Easley, D., Kleinberg, J., Tardos, E.: Trading networks with price-setting agents. Games Econ. Behav. **67**(1), 36–50 (2009)
3. Cheng, C.B., Chan, C.C.H., Lin, K.C.: Intelligent agents for e-marketplace: Negotiation with issue trade-offs by fuzzy inference systems. Decis. Support Syst. **42**(2), 626–638 (2006)
4. Fletcher, R.: Practical Methods of Optimization. John Wiley & Sons (2013)
5. Gupta, G.K.: Introduction to Data Mining with Case Studies. PHI Learning Pvt. Ltd. (2011)
6. Han, L., Hong, S.H.: In-house transactions in the real estate brokerage industry: matching outcome or strategic promotion?. In: Summer Real Estate Symposium. Monterey, California (2013)
7. Jiang, Z.Z., Fan, Z.P., Tan, C.Q., Yuan, Y.: A matching approach for one-shot multi-attribute exchanges with incomplete weight information in e-brokerage. Int. J. Innov. Comput. Inf. Control **7**(5), 2623–2636 (2011)
8. Jung, J.J., Jo, G.S.: Brokerage between buyer and seller agents using constraint satisfaction problem models. Decis. Support Syst. **28**(4), 293–304 (2000)
9. Li, X., Murata, T.: Priority based matchmaking method of buyers and suppliers in b2b e-marketplace using multi-objective optimization. In: Proceedings of the International Multi-Conference of Engineers and Computer Scientists, vol. 1 (2009)
10. Royo, A.S., Verdegay, J.L.: Methods for the construction of membership functions. Int. J. Intell. Syst. **14**(12), 1213–1230 (1999)
11. Saaty, T.L.: Exploring the interface between hierarchies, multiple objectives and fuzzy sets. Fuzzy Sets Syst. **1**(1), 57–68 (1978)
12. Zimmermann, H.J.: Fuzzy Sets, Decision Making, and Expert Systems, vol. 10. Springer Science & Business Media (2012)

Chapter 10
Multi-objective Nurse Rerostering Problem

Shih-Min Wu, Tenda Okimoto, Katsutoshi Hirayama and Katsumi Inoue

Abstract How to schedule a limited number of nurses in hospital wards staffed 24 h a day is important issue for the satisfactory patient care and potentially improve nurse retention. Nurse Scheduling Problem (NSP) is a combinatorial optimization problem, in which a set of nurses must be assigned into a limited set of working slots, subject to a given set of hard and soft constraints. Various sophisticated algorithms have been developed for solving a NSP. It is natural to consider the scheduled nurse's unexpected absences, e.g., illness, accident and injury. Nurse Rerostering Problem (NRP) is a dynamic NSP where the aim is to reschedule the current roster so that the number of changes of assignments between current and modified schedules is minimized. In this paper, the focus is laid on NRP with multiple criteria and the "egalitarianism" among nurses in a modified schedule. A formal framework of Multi-Objective Nurse Rerostering Problem (MO-NRP) is defined where the aim is to find trade-off solutions among "optimality" and "stability". Also, a novel solution criterion called an egalitarian solution for a MO-NRP is introduced.

Keywords Nurse rerostering problem · Multi-objective weighted constraint satisfaction problem · Pareto optimality · Stability · Egalitarianism

S.-M. Wu (✉) · T. Okimoto · K. Hirayama
Graduate School of Maritime Sciences, Kobe University,
Higashinadaku Fukaeminamimachi 5-1-1, Kobe 658-0022, Japan
e-mail: gsmin@stu.kobe-u.ac.jp

T. Okimoto
e-mail: tenda@maritime.kobe-u.ac.jp

K. Hirayama
e-mail: hirayama@maritime.kobe-u.ac.jp

K. Inoue
National Institute of Informatics, Chiyodaku Hitotsubashi 2-1-2,
Tokyo 101-8430, Japan
e-mail: inoue@nii.ac.jp

Q. Bai et al. (eds.), *Multi-agent and Complex Systems*, Studies in Computational Intelligence 670, DOI 10.1007/978-981-10-2564-8_10

10.1 Introduction

Nurse Scheduling Problem (NSP) [1, 6, 8, 12] is one of the widely investigated application problems in operations research (OR) and artificial intelligence (AI). It is well known that a NSP can be represented as an weighted constraint satisfaction problem (WCSP) [1, 14] where the aim is to find an assignment that satisfies all hard constraints and minimizes the sum of all violated costs of soft constraints. In order to provide the satisfactory patient care and potentially improve nurse retention, creating a good schedule for nurses is an important issue. However, since there are many constraints which must be satisfied, making an ideal schedule for both nurses and the hospital is intractable, and that is why the scheduler (e.g. head nurse in many cases) spends a lot of time to find a feasible schedule. Various sophisticated complete and incomplete algorithms have been introduced for solving a NSP in order to generate better nurse schedules and solve large-scale problems [2, 7, 9, 15].

Nurse Rerostering Problem (NRP) [16, 19, 21] is a dynamic NSP where the aim is to reschedule the current roster/schedule so that the number of changes of assignments between current and modified rosters/schedules is minimized. It is natural to consider the scheduled nurse's unexpected absences, e.g., illness, accident and injury of a nurse, after the scheduler created a roster with difficulty. When an absence is announced, the scheduler must find a nurse who can fill the vacancy of the absentee and the current schedule must be rebuilt as soon as possible. Most previous works on NRP have been investigated the stability of a modified schedule, i.e., a modified schedule should be similar to the previous one as much as possible.

The *egalitarianism* among nurses is an expected property of a NRP. Assume that the number of changes of all assignments in a modified schedule is small and it is also optimal (i.e. all hard constraints are satisfied and the sum of the violation costs of soft constraints is minimized). However, what happen if one nurse needs to change her assignments a lot in a modified schedule, while other nurses not. Clearly, the nurse who should change a lot complains about the modified schedule.

In this paper, the focus is laid on NRP with multiple criteria and the egalitarianism among nurses in a modified schedule. A formal framework for *Multi-Objective Nurse Rerostering Problem* (MO-NRP) is defined which is the extension of a mono-objective NRP. In this framework, the both stability and optimality are considered simultaneously. More specifically, MO-NRP is modeled by using the framework of a multi-objective WCSP [20] where the aim is to find an assignment that satisfies all hard constraints and minimizes the sum of violated costs of all objective functions.

Furthermore, a novel solution criterion called egalitarian solution for a MO-NRP is defined. In an egalitarian solution, the nurses share the changes of their shift works, i.e., minimize the maximal number of changes of assignments among nurses.

In a MO-NRP, since trade-offs exist among objectives, there does not generally exist an ideal assignment, which minimizes all objectives simultaneously. Thus, the optimal solutions of a MO-NRP is characterized by using the concept of *Pareto optimality*. An assignment is Pareto optimal if there does not exist another assignment that weakly improves all of the objectives. Solving a MO-NRP is to find Pareto front which is a set of cost vectors obtained by all Pareto optimal solutions (i.e. trade-off solutions among stability and optimality). MO-NRP can be represented using a graph called a constraint graph [22] in which nodes correspond to variables and each edge represents a constraint. In a MO-NRP, even if a constraint graph has the simplest tree structure, the size of Pareto front becomes exponential in the number of variables, i.e., all assignments are Pareto optimal solutions in the worst case.

The rest of the paper is organized as follows. In the next section, the formalizations of NRP and MO-WCSP are provided. Afterwards, the framework for MO-NRP is presented and the formal definition of an egalitarian solution for a MO-NRP is defined. Finally, we conclude this paper and give some future works.

10.2 Preliminaries

In this section, the models of nurse rerostering problem (NRP) and multi-objective weighted constraint satisfaction problem (MO-WCSP) are briefly described.

10.2.1 Nurse Rerostering Problem

Nurse Rerostering Problem (NRP) [16, 19, 21] is a dynamic nurse scheduling problem where the aim is to reschedule the current roster so that the number of changes of assignments between current and modified schedules is minimized, i.e., solving a NRP is to find a stable solution. In general, the constraints are dependent on the requirements of both nurses and hospitals. The following is the hard and soft constraints, which are frequently used in previous works. Note that some hard constraints are used as soft constraints and vice versa. It depends on the hospitals.

Hard Constraints

- H1: Prohibited working patterns must be avoided (e.g. one should not assign a nurse for 7 consecutive works and 3 consecutive night shifts).
- H2: In order to provide the satisfactory patient care, there exists the required number of nurses for each shift in a day (e.g., at least 3 nurses must be assigned to the morning and 2 nurses for evening and 1 nurse for night shifts).
- H3: For each nurse, the number of day-offs in a current schedule should not be less than that in a modified schedule.
- H4: Each newcomer should be assigned together with a skillful nurse, i.e., she has to work with a head nurse or a highly experienced nurse.
- H5: Nurses must rest at least 16 h between two consecutive shift works, e.g., in case a nurse is assigned to the night shift (0:00–8:00), morning (8:00–16:00) and evening shifts (16:00–24:00) should not be assigned.

Soft Constraints

S1: For each shift work (morning/evening/night), the required skill level of assigned nurses should be satisfied (e.g. for each shift work, at least one head nurse or one highly experienced nurse must be assigned).

S2: Day-offs of nurses in a current schedule should not be changed, i.e., the scheduled day-offs after modification must be same as much as possible.

S3: Requests of nurses (e.g. the preferred working patterns and specially the day-off requests) should be satisfied as much as possible.

Objective: Minimize the number of changes of shift works between current and modified schedules, i.e., the aim is to find a stable solution.

10.2.2 Multi-objective WCSP

Multi-Objective Weighted Constraint Satisfaction Problem (MO-WCSP) [20] is the extension of a mono-objective WCSP [1, 14] where the aim is to find an assignment that satisfies all hard constraints and minimizes the sum of all violated costs of soft constraints. Let k be the number of objectives. MO-WCSP is defined by a tuple MO-WCSP = $\langle X, D, C, S, \Phi \rangle$, where $X = \{x_1, x_2, ..., x_n\}$ is a set of variables, $D = \{d_1, d_2, ..., d_m\}$ is a set of domains, $C = \{C^1, C^2, ..., C^k\}$ is a set of hard and soft constraints, $S = \{S^1, S^2, ..., S^k\}$ is a set of valuation structures, and $\Phi = \{\phi^1, \phi^2, ..., \phi^k\}$ is a set of multi-objective functions. For each objective i $(1 \leq i \leq k)$, $C^i = C^i_h \cup C^i_s$ is the union of hard and soft constraints, where C^i_h is a set of hard constraints and C^i_s shows a set of soft constraints, $S^i = (E^i, \sum, <)$ is the valuation structure, where $E^i = \mathbb{N} \cup \{\infty\}$, $\sum$ is the standard sum over $\mathbb{N}$ and all elements of E are ordered by the operator $<$, and $\phi^i : C^i \rightarrow E^i$ is a cost function. Let A be an assignment to all variables. For an objective i, the valuation of A for constraint $c \in C^i$ is defined as:

$$\phi^i(A, c) = \begin{cases} 0 & c \in C^i_h \ is\ satisfied\ by\ A, \\ \infty & c \in C^i_h \ is\ violated\ by\ A, \\ \phi^i(A, c) & c \in C^i_s, \end{cases}$$

and the overall valuation of A is given by

$$\phi^i(A) = \sum_{c \in C^i} \phi^i(A, c).$$

Then, the sum of the violation costs of all cost functions for k objectives is defined by a cost vector, denoted

$$\Phi(A) = (\phi^1(A), \phi^2(A), ..., \phi^k(A)).$$

Finding an assignment that minimizes all objective functions simultaneously is ideal. However, in general, since trade-offs exist among objectives, there does not exist such an ideal assignment. Therefore, the "optimal" solution of a MO-WCSP is characterized by using the concept of Pareto optimality. This problem can be represented using a graph (called a constraint graph [22]), in which each node corresponds to a variable and each edge represents a constraint. In a MO-WCSP, even if a constraint graph has the simplest tree structure, the number of Pareto optimal solutions is often exponential in the number of variables in the worst case.

Definition 1 (*Dominance*) For a MO-WCSP, two cost vectors $\Phi(A) = (\phi^1(A), \phi^2(A), ..., \phi^k(A))$ and $\Phi(A') = (\phi^1(A'), \phi^2(A'), ..., \phi^k(A'))$, we call that $\Phi(A)$ dominates $\Phi(A')$, denoted by $\Phi(A) \prec \Phi(A')$, iff $\Phi(A)$ is partially less than $\Phi(A')$, i.e., it holds

- $\phi^i(A) \leq \phi^i(A')$ for all objectives i, and
- there exists at least one objective i', such that $\phi^{i'}(A) < \phi^{i'}(A')$.

Definition 2 (*Pareto optimal solution*) For a MO-WCSP, an assignment A is said to be Pareto optimal solution, iff there does not exist another assignment A', such that $\Phi(A') \prec \Phi(A)$.

Definition 3 (*Pareto Front*) For a MO-WCSP, a set of cost vectors obtained by Pareto optimal solutions is said to be Pareto front. Solving a MO-WCSP is to find Pareto front.

Example 1 (*MO-WCSP*) Consider the complete graph (i.e. each node has constraints with all other nodes) of a bi-objective WCSP with three variables x_1, x_2 and x_3 (see Fig. 10.1). Each node represents a variable and each edge corresponds to a constraint between two variables. Each variable takes its value from finite, discrete domain $\{0, 1\}$. The table shows the cost vectors for each constraint. For example, for the constraint between x_1 and x_3 (middle in the table), in case x_1 takes the value 0 and x_3 takes 1, the obtained cost vector is $(0, 3)$, i.e., the violation cost is 0 for objective 1 and 3 for objective 2. The cost ∞ in the table means that it violates a hard constraint. Pareto optimal solutions of this problem are $\{\{(x_1, 0), (x_2, 1), (x_3, 0)\}, \{(x_1, 0), (x_2, 1), (x_3, 1)\}\}$ and the obtained Pareto front is $\{(3, 6), (4, 5)\}$.

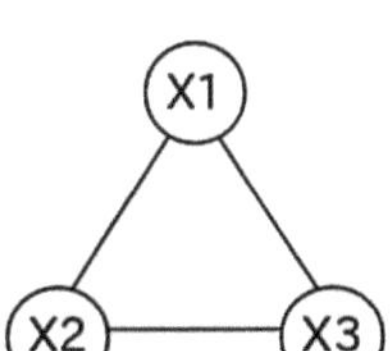

x_1	x_2	*cost*	x_1	x_3	*cost*	x_2	x_3	*cost*
0	0	(∞,0)	0	0	(3,1)	0	0	(1,3)
0	1	(0,1)	0	1	(0,3)	0	1	(2,∞)
1	0	(∞,1)	1	0	(3,2)	1	0	(0,4)
1	1	(∞,0)	1	1	(2,1)	1	1	(4,1)

Fig. 10.1 Example of a bi-objective WCSP with three variables x_1, x_2 and x_3. Each node represents a variable and each edge corresponds to a constraint between two variables. Each variable takes its value from discrete domain $\{0, 1\}$. Table shows the cost vectors for each constraint. Pareto optimal solutions are $\{\{(x_1, 0), (x_2, 1), (x_3, 0)\}, \{(x_1, 0), (x_2, 1), (x_3, 1)\}\}$ and Pareto front is $\{(3, 6), (4, 5)\}$

10.3 Multi-objective Nurse Rerostering Problem

In order to consider minimizing the number of constraint violations (optimality) and the number of changes of assignments (stability) simultaneously in a NRP, i.e., NRP with multiple criteria, a Multi-Objective Nurse Rerostering Problem (MO-NRP) is formalized. Moreover, a novel solution criterion called an egalitarian solution for a MO-NRP is defined. First, let us describe the following basic terms for a MO-NRP.

- $N = \{1, ..., n\}$ is a set of ID-numbers for nurses.
- $M = \{1, ..., m\}$ is a set of days in a scheduling period.
- $X = \{x_{11}, ..., x_{nm}\}$ is a set of variables.
- $W = \{o, m, e, n\}$ is a set of shift works, where o = {day-off}, m = {morning} (8:00–16:00), e = {evening} (16:00–24:00) and n = {night} (0:00–8:00).
- $L = \{l_1, ..., l_5\}$ is a set of skill levels of nurses where l_1 = {head nurse}, l_2 = {highly experienced}, l_3 = {experienced (i.e. more than 3 years)}, l_4 = {few years experience (i.e. 1-2 years)} and l_5 = {newcomer}.
- $\alpha_l : N \to L$ is a mapping which provides the skill level of a nurse, e.g., for a head nurse $i \in N$, her skill level can be obtained by $\alpha_l(i) = l_1$.

A $(n \times m)$-table is said to be a master schedule and is denoted as $MS_{current}$ for a current schedule and MS_{mod} for a modified schedule after unexpected absences of a nurse. One can see that $MS_{current}$ is a solution of NSP and MS_{mod} is that of NRP.

Definition 4 (*Stability*) For two master schedules $MS_{current}$ and MS_{mod}, each $w_{ij} \in W$ in $MS_{current}$ and each $w'_{ij} \in W'$ in MS_{mod}, and a non-negative integer r, MS_{mod} is said to be *r-stable*, iff the sum of the changes of assignments is bounded by r, i.e.,

$$\sum_{i,j} g(w_{ij}, w'_{ij}) \leq r, \ where \ g(w_{ij}, w'_{ij}) = \begin{cases} 0 & w_{ij} = w'_{ij}, \\ 1 & otherwise. \end{cases}$$

Example 2 Consider a master schedule for a week of 7 nurses. Table 10.1 (left) represents the current master schedule $MS_{current}$ which satisfies all hard constraints provided in Sect. 10.2, i.e. hard constraints from H1 to H5. Assume that nurse n_5 has an unexpected absence on Monday and cannot work her morning shift work m. Table 10.1 (right) shows a modified master schedule MS_{mod}. The morning shift of n_5 on Monday has been changed from m to absence in MS_{mod} (denoted by ╱). From the hard constraint H2, i.e., at least 3 nurses must be assigned to the morning shift and 2 for evening and 1 for night shifts, nurse n_1 works the morning shift work m instead of n_5 in MS_{mod}. In order to satisfy all hard constraints, nurse n_1 changes her shift works (i.e. evening shifts e) on Friday, Saturday and Sunday in $MS_{current}$ to night shift n on Friday, day-off o on Saturday and morning shift m on Sunday in MS_{mod}. Also, nurse n_5 changes her night shift n on Friday, day-off o on Saturday and morning shift m on Sunday in $MS_{current}$ to evening shifts e on these three days in MS_{mod}. Since the number of changes of shift works between $MS_{current}$ and MS_{mod} is 8 (including the absence of nurse n_5 on Monday), MS_{mod} is r = 8-stable.

Table 10.1 Example of $MS_{current}$ for a week of 7 nurses (left) and a modified schedule MS_{mod} (right). Nurse n_5 had an unexpected absence on Monday (denoted by /). Red fonts show the modified shift works. Nurse n_1 changes her shift works (i.e. evening shifts e) on Friday, Saturday and Sunday in $MS_{current}$ to night shift n on Friday, day-off o on Saturday and morning shift m on Sunday in MS_{mod}. Nurse n_5 changes her night shift n on Friday, day-off o on Saturday and morning shift m on Sunday in $MS_{current}$ to evening shifts e on these three days in MS_{mod}. The MS_{mod} is $r = 8$-stable

$MS_{current}$							
Nurse (Level)	M	T	W	T	F	S	S
n_1 (l_1)	o	m	m	m	e	e	e
n_2 (l_2)	e	e	n	o	m	m	m
n_3 (l_3)	m	m	m	e	e	n	o
n_4 (l_3)	m	e	e	n	o	m	n
n_5 (l_4)	m	m	e	e	n	o	m
n_6 (l_4)	n	n	o	m	m	e	e
n_7 (l_5)	e	o	m	m	m	m	m

MS_{mod}							
Nurse (Level)	M	T	W	T	F	S	S
n_1 (l_1)	m	m	m	m	n	o	m
n_2 (l_2)	e	e	n	o	m	m	m
n_3 (l_3)	m	m	m	e	e	n	o
n_4 (l_3)	m	e	e	n	o	m	n
n_5 (l_4)	/	m	e	e	e	e	e
n_6 (l_4)	n	n	o	m	m	e	e
n_7 (l_5)	e	o	m	m	m	m	m

The framework for MO-NRP is defined as follows.

Definition 5 (MO-NRP) A multi-objective nurse rerostering problem is a tuple

$$\text{MO-NRP} = \langle X, W, L, C, S, MS_{current}, \Phi \rangle,$$

where X is a set of variables, W is a set of domains, L is a set of skill levels, C and S are same as a MO-WCSP, $MS_{current}$ is the current schedule, $\Phi = \{\phi^{opt}, \phi^{stable}\}$ is a set of cost functions where ϕ^{opt} is a cost function for optimality and ϕ^{stable} is that for stability. For a value assignment A to all variables, the sum of the violation costs and the number of the changes of assignments are given by a vector $\Phi(A) = (\phi^{opt}(A), \phi^{stable}(A))$. Solving a MO-NRP is to find Pareto optimal solutions so that

1. all hard constraints are satisfied,
2. the sum of the violation costs of soft constraints is minimized (i.e. optimality),
3. the number of the changes of assignments is minimized (i.e. stability).

In previous works on NRP, the aim is to find an assignment so that the number of the changes of assignments between current and modified schedules is minimized, i.e., solving a NRP is to find a stable solution. On the other hand, in a MO-NRP, bi-objectives are considered simultaneously, namely optimality and stability. In this framework, one can easily define several objective functions (i.e. $\phi^{opt_1}, \phi^{opt_2}, \ldots, \phi^{opt_p}$) instead of only one objective function ϕ^{opt} by considering each soft constraint as an objective function. For the simplicity, this paper defines ϕ^{opt} for optimality like classic NSP. Such simplification can be done by aggregating all objective functions which is called an AOF technique [17] (or in other words, linear sum and scalarization methods). Note that this technique can be utilized among objective functions for optimality and not for objective functions for optimality and stability, i.e., it makes no sense to aggregate the costs and the number of changes.

Definition 6 (s-*vector*) Let MS_{mod} be a modified master schedule. For a nurse i ($1 \leq i \leq n$), let s_i be the number of changes of assignments from a current master schedule to MS_{mod}. The number of changes of assignments for all nurses is said to be a s-vector w.r.t. MS_{mod} and denoted by $v_s = (s_1, ..., s_n)$.

Definition 7 (*Equivalence*) For two s-vectors $v_s = (s_1, ..., s_n)$ and $v_{s'} = (s'_1, ..., s'_n)$ w.r.t. MS_{mod}, v_s and $v_{s'}$ are said to be equivalent, iff it holds

$$\sum_{i=1}^{n} s_i = \sum_{i=1}^{n} s'_i$$

Let V_s be a set of equivalent s-vectors w.r.t. MS_{mod} and $\preceq_{lex}$ be the total preorder over V_s defined $\forall v_s, v_{s'} \in V_s$ as $v_s \preceq_{lex} v_{s'}$ if and only if lexically reordered v_s precedes lexically reordered $v_{s'}$. For example, let $v_s = (4, 1, 3, 2, 2)$ and $v_{s'} = (4, 0, 3, 2, 3)$ be two equivalent s-vectors (i.e. $\sum_{i=1}^{5} s_i = 4 + 1 + 3 + 2 + 2 = 12 = 4 + 0 + 3 + 2 + 3 = \sum_{i=1}^{5} s'_i$). The corresponding reordered vectors are $\overline{v_s} = (4, 3, 2, 2, 1)$ and $\overline{v_{s'}} = (4, 3, 3, 2, 0)$. Compare the 1st components of $\overline{v_s}$ and $\overline{v_{s'}}$. In case they are same, the 2nd components are compared. Continue to compare until one of two components is smaller than the another one. In this example, for the 3rd components, since 2 of $\overline{v_s}$ is smaller than 3 of $\overline{v'_s}$, the vector v_s is lexically smaller than v'_s (i.e. $v_s \preceq_{lex} v_{s'}$).

Definition 8 (*Egalitarianism*) For a modified master schedule MS_{mod} and a s-vector v_s w.r.t. MS_{mod}, v_s is said to be an egalitarian solution of MS_{mod}, iff there does not exist another equivalent s-vector $v_{s'}$ w.r.t. MS_{mod}, such that

$$v_{s'} \preceq_{lex} v_s,$$

i.e., minimizing the maximal number of changes among nurses.

Example 3 Consider the master schedules in Table 10.2. The MS_{mod} is the master schedule presented in Table 10.1 and the MS'_{mod} shows an alternative master schedule. The s-vectors v_s w.r.t. MS_{mod} and $v_{s'}$ w.r.t. MS'_{mod} are

$$v_s = (4, 0, 0, 0, 4, 0, 0), \quad v_{s'} = (2, 0, 1, 1, 4, 0, 0).$$

Since the number of changes of assignments is 8, the MS'_{mod} is also $r = 8$-stable, i.e., v_s and $v_{s'}$ are equivalent. The lexically reordered vectors of v_s and $v_{s'}$ are

$$\overline{v_s} = (4, 4, 0, 0, 0, 0, 0), \quad \overline{v_{s'}} = (4, 2, 1, 1, 0, 0, 0).$$

Table 10.2 MS_{mod} (left) is the modified schedule used in Example 2. MS'_{mod} (right) is an alternative modified schedule which is also 8-stable like MS_{mod}. The MS'_{mod} is more egalitarian than MS_{mod}

MS_{mod}							
Nurse (Level)	M	T	W	T	F	S	S
n_1 (l_1)	**m**	*m*	*m*	*m*	**n**	**o**	**m**
n_2 (l_2)	*e*	*e*	*n*	*o*	*m*	*m*	*m*
n_3 (l_3)	*m*	*m*	*m*	*e*	*e*	*n*	*o*
n_4 (l_3)	*m*	*e*	*e*	*n*	*o*	*m*	*n*
n_5 (l_4)	/	*m*	*e*	*e*	**e**	**e**	**e**
n_6 (l_4)	*n*	*n*	*o*	*m*	*m*	*e*	*e*
n_7 (l_5)	*e*	*o*	*m*	*m*	*m*	*m*	*m*

MS'_{mod}							
Nurse (Level)	M	T	W	T	F	S	S
n_1 (l_1)	**m**	*m*	*m*	*m*	*e*	**o**	*e*
n_2 (l_2)	*e*	*e*	*n*	*o*	*m*	*m*	*m*
n_3 (l_3)	*m*	*m*	*m*	*e*	**n**	*n*	*o*
n_4 (l_3)	*m*	*e*	*e*	*n*	*o*	*m*	**m**
n_5 (l_4)	/	*m*	*e*	*e*	**e**	**e**	**n**
n_6 (l_4)	*n*	*n*	*o*	*m*	*m*	*e*	*e*
n_7 (l_5)	*e*	*o*	*m*	*m*	*m*	*m*	*m*

The s-vector $v_{s'}$ w.r.t. MS'_{mod} is more egalitarian than v_s, i.e., $v_{s'} \preceq v_s$. Compared to MS_{mod}, four nurses (i.e. n_1, n_3, n_4 and n_5) share the changes of their shift works in MS'_{mod}, while only two nurses (i.e. n_1 and n_5) changes their assignments in MS_{mod}.

10.4 Experiments

In this section, an egalitarian solution for a MO-NRP is computed by using the Lp solver (Lp solve IDE 5.5.2.0). In the experiments, a master schedule is created. Then, the modified schedules are generated by absenting any nurse in the master schedule. The experimental setting is as follows.

- Period: one week (from Monday M to Sunday S).
- The number of nurses: 7 (n_1, n_2,..., n_7).
- Hard constraints
  - One should not assign a nurse for 7 consecutive works and 3 consecutive night shifts (H1).
  - At least 2 nurses must be assigned to the morning and 2 nurses for evening and 1 nurse for night shifts (H2).
  - For each nurse, the number of day-offs in a current schedule should be same in a modified schedule (H3).
  - Nurses must rest at least 16 hours between two consecutive shift works (H5).
- Soft constraint: In each day, at least one head nurse or one highly experienced nurse must be assigned (S1).
- Objective 1: Minimize the number of violations of the soft constraint.
- Objective 2: Minimize the number of changes between the master and modified schedules.

The following shows the Lp program we used in the experiments in order to compute the number of changes of assignments (i.e. objective 2). Let $N = \{i \mid 1, ..., n\}$ be a set of nurses, $M = \{j \mid 1 \leq j \leq m\}$ be a set of days in a schedule period, and $W = \{k \mid 1 \leq k \leq 4\}$ be a set of shift works, where 1=day-off, 2=morning, 3=evening and 4=night.

$$Minimize\ 1 - \sum x_{ijk} \tag{10.1}$$

subject to

$$\sum_{k \in W} x_{ijk} = 1 \tag{10.2}$$

$$x_{ijk} + x_{i(j+1)k} + x_{i(j+2)k} + x_{i(j+3)k} + x_{i(j+4)k} + x_{i(j+5)k} + x_{i(j+6)k} \leq 6 \tag{10.3}$$

$$x_{ij4} + x_{i(j+1)4} + x_{i(j+2)4} \leq 2 \tag{10.4}$$

$$\sum_{i \in N} x_{ij2} \geq 2, \quad \sum_{i \in N} x_{ij3} \geq 2, \quad \sum_{i \in N} x_{ij4} \geq 1 \tag{10.5}$$

$$x_{i'j'1} = 1 \tag{10.6}$$

(1) represents the objective function 2, i.e., minimize the number of changes between the master and modified schedules. (2) is the constraint that no one can work several shifts in a day, e.g., morning and evening in the same day. (3) and (4) represent the forbidden shift patters for H1, i.e., 7 consecutive works and 3 consecutive night shifts. (5) is the constraint for H2 and (6) shows that a nurse i' has unexpected absent on a day j'.

In the experiments, we aggregate the objective function 1 and 2 and find the optimal solution which minimizes the sum of the costs. Table 10.3 shows a master schedule which satisfies all hard constraints, and the number of violations of the soft constraint is zero. Table 10.4 represents two modified schedules we computed. The both schedules satisfy all hard constraints and the number of violations of the soft constraint is zero.

The s-vectors v_s w.r.t. MS_{mod} and $v_{s'}$ w.r.t. MS'_{mod} are

$$v_s = (0, 0, 0, 2, 0, 2, 2), \quad v_{s'} = (1, 2, 0, 1, 0, 2, 0).$$

Table 10.3 Master schedule for one week with 7 nurses. This schedule satisfies all hard constraints and the number of violations of the soft constraint is zero

$MS_{current}$: Master schedule							
Nurse Level	M	T	W	T	F	S	S
n_1 (l_1)	*m*	*o*	*n*	*e*	*o*	*m*	*m*
n_2 (l_2)	*o*	*m*	*m*	*o*	*m*	*m*	*m*
n_3 (l_3)	*e*	*e*	*o*	*m*	*m*	*o*	*e*
n_4 (l_3)	*n*	*o*	*m*	*m*	*e*	*o*	*e*
n_5 (l_4)	*m*	*m*	*e*	*o*	*e*	*e*	*o*
n_6 (l_4)	*e*	*e*	*o*	*n*	*n*	*e*	*o*
n_7 (l_5)	*o*	*n*	*e*	*e*	*o*	*n*	*n*

Table 10.4 Modified schedules where the nurse n_6 has unexpected absent

MS_{mod}							
Nurse (Level)	M	T	W	T	F	S	S
n_1 (l_1)	*m*	*o*	*n*	*e*	*o*	*m*	*m*
n_2 (l_2)	*o*	*m*	*m*	*o*	*m*	*m*	*m*
n_3 (l_3)	*e*	*e*	*o*	*m*	*m*	*o*	*e*
n_4 (l_3)	**e**	*o*	*m*	*m*	*e*	*o*	**n**
n_5 (l_4)	*m*	*m*	*e*	*o*	*e*	*e*	*o*
n_6 (l_4)	/	*e*	*o*	*n*	*n*	*e*	**e**
n_7 (l_5)	**n**	*n*	*e*	*e*	*o*	*n*	**o**

MS'_{mod}							
Nurse (Level)	M	T	W	T	F	S	S
n_1 (l_1)	**e**	*o*	*n*	*e*	*o*	*m*	*m*
n_2 (l_2)	**m**	*m*	*m*	*o*	*m*	*m*	**o**
n_3 (l_3)	*e*	*e*	*o*	*m*	*m*	*o*	*e*
n_4 (l_3)	*n*	*o*	*m*	*m*	*e*	*o*	**m**
n_5 (l_4)	*m*	*m*	*e*	*o*	*e*	*e*	*o*
n_6 (l_4)	/	*e*	*o*	*n*	*n*	*e*	**e**
n_7 (l_5)	*o*	*n*	*e*	*e*	*o*	*n*	*n*

Since the number of changes of assignments is 6 in both schedules (i.e. they are $r = 6$-stable), v_s and $v_{s'}$ are equivalent. Their lexically reordered vectors are

$$\overline{v_s} = (2, 2, 2, 0, 0, 0, 0), \quad \overline{v_{s'}} = (2, 2, 1, 1, 0, 0, 0).$$

The s-vector $v_{s'}$ w.r.t. MS'_{mod} is an egalitarian solution for this problem instance.

10.5 Related Work

Compared to NSP, there exists few works on NRP. Moz et al. [19] proposed two integer multicommodity flow models for a NRP. The first one is a directed multilevel acyclic network based model where the aim is to optimize an integer multicommodity flow in a multi-level network by adding some constraints. The other one is the extension of the first one which is an aggregation based model (i.e. aggregate the nodes of this network). They empirically showed that the second model outperforms the first, both the solution quality and runtime. Hattori et al. [11] formalized a dynamic NSP by using the framework of dynamic weighted MaxCSP which can effectively deal with dynamic changes to a problem. They introduced provisional constraints which allow variables to keep the same values so that one can obtain stable solutions that are close to previous ones. Pato et al. [21] worked on a utopic Pareto genetic heuristic which considers the trade-offs between two objectives, i.e., (i) minimize the gap between the number of scheduled duties and the number of duties each nurse should perform during the period, and (ii) minimize dissimilarity regarding the previously announced roster for the same period. Maenhout et al. [16] developed an evolutionary meta-heuristic which revises and re-optimizes a schedule for a set of heterogeneous nurses. Compared to these existing works, this paper focuses on NRP with multiple criteria and also the egalitarianism among nurses.

There exists very limited work on NSP with multiple criteria [5, 8]. The goal programming is the most widely used method where the aim is to find a solution which is as close as possible to each of the objectives in the order of the given priorities [3]. Others are the well-known tabu search based approach [5], Pareto simulated annealing approach based on the scalarization [13], modified harmony search [2] and adaptive neighborhood search [15]. Compare to these existing works, this paper focuses on a dynamic multi-objective NSP (i.e. MO-NRP).

NRP can be an application problem of Minimal Perturbation Problem (MPP) [10, 24] which is a dynamic CSP where the aim is to find a solution that minimizes a given distance function. The distance function measures the number of changing variables. Minimizing perturbations results in minimizing the number of changes in the assignment. Solving a MPP is finding a stable solution like NRP. Compared to MPP, this paper focuses on MO-NRP and also the egalitarianism among nurses.

10.6 Conclusion

In order to provide satisfactory patient care and potentially improve nurse retention, creating a good schedule for nurses and hospitals is important issue. NSP is a combinatorial optimization problem, in which a set of nurses must be assigned into a limited set of working slots, subject to a given set of hard and soft constraints. It is natural to consider the scheduled nurse's unexpected absence. NRP is a dynamic NSP where the aim is to reschedule the current roster after the unexpected absence of a nurse so that (i) all hard constraints are satisfied and (ii) the number of the changes of assignments between current and modified schedules is minimized. Most previous works on NRP focused on the stability, i.e., the new schedule should be similar to the current one as much as possible. The contribution of this paper is twofold:

- A formal framework of Multi-Objective Nurse Rerostering Problem (MO-NRP) is first defined by using the framework of a multi-objective weighted constraint satisfaction problem. The aim of a MO-NRP is to find trade-off solutions among "optimality" and "stability" of a modified schedule.

- A novel solution criterion in a MO-NRP is introduced, namely an egalitarian solution. By considering the egalitarianism among nurses, the following situation can be avoided; some nurses change their assignments a lot, while others not, i.e., in an egalitarian solution, the changes of assignments are shared among nurses.

As a perspective for further research, we intend to apply our approach to some real problems and analyze the trade-off solutions for a MO-NRP. More specifically, for existing NSP benchmarks in INRC-II (the second international nurse rostering competition), we model them as MO-NRP by assuming all soft constraints as objective functions and find an egalitarian solution. In order to solve the problems, we will use existing SAT/ASP solvers [4, 23]. We are also interested in multi-objective setting and egalitarian solutions in sport and transport timetables [18].

References

1. Abdennadher, S., Schlenker, H.: Nurse Scheduling using constraint logic programming. In: Proceedings of the 16th National Conference on Artificial Intelligence, pp. 838–843 (1999)
2. Awadallah, M., Khader, A., Al-Betar, M., Bolaji, A.: Nurse rostering using modified harmony search algorithm. In: Proceedings of the 6th International Conference on Bio-Inspired Computing: Theories and Applications, pp. 58–63 (2011)
3. Azaieza05, M., Sharif, S.A.: A 0-1 goal programming model for nurse scheduling. Comput. Oper. Res., **31**, 491–507 (2005)
4. Banbara, M., Soh, T., Tamura, N., Inoue, K., Schaub, T.: Answer set programming as a modeling language for course timetabling. TPLP **13**(4–5), 783–798 (2013)
5. Burke, E., Bb, N.N., Rostering, T.N., Causmaecker, P.D.: A multi criteria meta-heuristic approach to nurse rostering. In: Proceedings of Congress on Evolutionary Computation, pp. 1197–1202 (2002)
6. Burke, E., Causmaecker, P., Berghe, G., Landeghem, H.: The state of the art of nurse rostering. J. Sched. **7**(6), 441–499 (2004)
7. Burke, E., Curtois, T.: New approaches to nurse rostering benchmark instances. Eur. J. Oper. Res. **237**(1), 71–81 (2014)
8. Burke, E., Li, J., Qu, R.: A pareto-based search methodology for multi-objective nurse scheduling. Ann. OR **196**(1), 91–109 (2012)
9. Causmaecker, P., Berghe, G.: A categorisation of nurse rostering problems. J. Sched. **14**(1), 3–16 (2011)
10. Fukunaga, A.: An improved search algorithm for min-perturbation. In: Proceedings of the 19th International Conference on Principles and Practice of Constraint Programming, pp. 331–339 (2013)
11. Hattori, H., Ito, T., Ozono, T., Shintani, T.: A nurse scheduling system based on dynamic constraint satisfaction problem. In: Proceedings of the 18th International Conference on Industrial and Engineering Applications of Artificial Intelligence and Expert Systems (IEA/AIE-2005), pp. 799–808 (2005)
12. Ikegami, A., Niwa, A.: A subproblem-centric model and approach to the nurse scheduling problem. Math. Program. **97**(3), 517–541 (2003)
13. Jaszkiewicz, A.: Foundations of computing and decision sciences. Found. Comput. Decis. Sci. **22**, 169–184 (1997)
14. Larrosa, J., Schiex, T.: Solving weighted csp by maintaining arc consistency. Artif. Intell. **159**(1–2), 1–26 (2004)
15. Lü, Z., Hao, J.: Adaptive neighborhood search for nurse rostering. Eur. J. Oper. Res. **218**(3), 865–876 (2012)

16. Maenhout, B., Vanhoucke, M.: An evolutionary approach for the nurse rerostering problem. Comput. OR **38**(10), 1400–1411 (2011)
17. Miettinen, K.: Nonlinear Multiobjective Optimization. Kluwer Academic Publishers, Boston (1999)
18. Mirhassani, S., Habibi, F.: Solution approaches to the course timetabling problem. AI Review. **39**(2), 133–149 (2013)
19. Moz, M., Pato, M.: Solving the problem of rerostering nurse schedules with hard constraints: New multicommodity flow models. Ann. OR **128**(1–4), 179–197 (2004)
20. Okimoto, T., Ribeiro, T., Clement, M., Inoue, K.: Modeling and algorithm for dynamic multi-objective weighted constraint satisfaction problem. In: Proceedings of the 6th International Conference on Agents and Artificial Intelligence, pp. 420–427 (2014)
21. Pato, M., Moz, M.: Solving a bi-objective nurse rerostering problem by using a utopic pareto genetic heuristic. J. Heuristics **14**(4), 359–374 (2008)
22. Schiex, T., Fargier, H., Verfaillie, G.: Valued constraint satisfaction problems: hard and easy problems. In: Proceedings of the 14th International Joint Conference on Artificial Intelligence, pp. 631–639 (1995)
23. Soh, T., Tamura, N., Banbara, M.: Scarab: a rapid prototyping tool for sat-based constraint programming systems. In: Theory and Applications of Satisfiability Testing—SAT 2013—16th International Conference, Helsinki, Finland, 8–12 Jul 2013. Proceedings, pp. 429–436 (2013)
24. Zivan, R., Grubshtein, A., Meisels, A.: Hybrid search for minimal perturbation in dynamic csps. Constraints **16**(3), 228–249 (2011)

Chapter 11
Preference Aware Influence Maximization

Chang Jiang, Weihua Li, Quan Bai and Minjie Zhang

Abstract With the development of social network, online marketing has become more popular and developed in an unprecedented scale. Viral marketing propagates influence through 'word-of-mouth' effect. As for development of viral marketing, it is critical to select a set of influential users in the network to propagate influence as much as possible with limited resources. In this chapter, we proposed a model called Preference-based Trust Independent Cascade Model. Based on the experimental results, the Preference-based Trust Independent Cascade Model is able to obtain better results than some traditional models. Comparing with other existing methods, such as trust-only approach and random selection approach, the proposed Preference-based Trust Independent Cascade Model considers both user preference and trust connectivity.

Keywords Influence maximization · Preference-aware · Influence propagation

11.1 Introduction

With the development of on-line social networks and e-Marketplace, it is essential to understand how to propagate and maximize influence to users with limited resources, namely, influence maximization (IM) [3]. IM has become a popular topic for researchers in both computer science and business. The solution for IM problem is NP-hard. Hence, some approximation approaches are considered as a replace-

C. Jiang (✉) · W. Li · Q. Bai
Auckland University of Technology, Auckland, New Zealand
e-mail: rky9795@autuni.ac.nz

W. Li
e-mail: weihua.li@aut.ac.nz

Q. Bai
e-mail: quan.bai@aut.ac.nz

M. Zhang
University of Wollongong, Wollongong, Australia
e-mail: minjie@uow.edu.au

Q. Bai et al. (eds.), *Multi-agent and Complex Systems*, Studies in Computational Intelligence 670, DOI 10.1007/978-981-10-2564-8_11

ment. Most of the existing approaches, such as basic diffusion models [7], only consider about propagation links/channels in IM (called trust connectivity (TC) in this chapter). For instance, assuming there are two users following each other on Twitter. In this situation, the two users have TC between each other. By considering TC only, IM will be treated as a simplified probabilistic problem. Furthermore, influence probabilities are predefined and static, which is not suitable for dynamic environments in many real-world applications. Meanwhile, in a social network, users' preferences to particular items also play an important role in IM [13]. Unfortunately, this factor was not considered in most of the existing IM approaches.

Motivated by the prosperous development of the viral marketing and in order to cover the aforementioned limitations in existing approaches, in this chapter, a Preference-based Trust Independent Cascade (PTIC) Model is proposed. The PTIC Model takes into account not only TC but also user preference (UP). The influence probability in PTIC Model is computed based on UP and TC. So that, hub users in a social network, who are interested in promoted items, can be selected as influential users. Through this way, the effectiveness and efficiency of IM can be greatly improved. Experimental results also demonstrate that the PTIC Model has a better performance compared with some existing classic models.

The reminder of this chapter is organized as follows. In Sect. 11.2, some related work has been reviewed. Section 11.3 presents the problem description and formal definitions related to the PTIC Model. In Sect. 11.4, the framework of the PTIC Model is described in details. Experiments are presented in Sect. 11.5. Finally, the chapter is concluded in Sect. 11.6.

11.2 Related Work

IM is first proposed by Domingos and Richardson as a probabilistic problem [4, 11]. Kempe et al. are the earliest to research IM as a discrete optimization problem [7]. They have demonstrated that the solution for IM problem is NP-hard. Finding the solution for a NP-hard problem is very difficult. Thus, instead of finding the solution for a NP-hard problem, approximation approaches are better replacement with guarantee [1, 2, 13].

There are two fundamental models for influence diffusion, i.e., the Linear Threshold (LT) Model and the Independent Cascade (IC) Model [3, 12]. Many studies are developed based on these two models. For example, Chen et al. proposed the first Scalable Heuristic Algorithm based on LT Model and a new Degree Discount Heuristics based on IC Model [3, 7, 12]. However, network analysis approaches, such as, community partition, are not involved in the scalable heuristic algorithm. Community partition can improve the efficiency and effectiveness of IM in a further step. As for the degree discount heuristics, it is efficient but it is sensitive when the influence probabilities are small only. In the real world, it is difficult to influence users with a small probability.

UP also plays an important role in IM. However, many existing preference analysis approaches are exploited for recommender systems [9, 10]. By considering UP, users in a social network can be clustered into different community partitions based on their common preferences. Furthermore, hierarchical communities can be identified in a network based on hierarchical clustering (HC), also called hierarchical clustering analysis.

A representative algorithm of HC is the Shrink algorithm [6]. Shrink is an unsupervised clustering algorithm. It takes advantage of the integration of the modularity optimization approaches and the density-based clustering. The remarkable merits of the Shrink algorithm are that it can detect not only clusters, but also the structure of the network, i.e., the hubs and the outliers. Another HC algorithm is a community partition method based on node similarity proposed by Ying el al. [8]. In the algorithm, each node is initialized as a community in the network at the beginning. Then, communities are merged iteratively based on neighbourhood similarity. Ying et al.'s approach has lower computational complexity and has been applied for many types of networks, indicating that this algorithm is effective and efficient for community detection.

Based on reviews above, the limitations of the existing approaches have been summarized. First of all, some classic diffusion models, i.e., IC Model and LT Model only consider TC, which result in that IM is treated as a simplified probabilistic problem. In addition, the influence probabilities are predefined and static. Secondly, UP plays a critical part in IM, since users with common preferences can easier influence each other. However, this factor was not taken into consideration in most of the existing approaches.

UP and TC are involved in the PTIC Model, and the influence probability is computed based on both of them. Hence, hub users who are interested in the promoted item, will be selected as influential users. In this way, effectiveness and efficiency of IM can be improved significantly.

11.3 Problem Description and Formal Definitions

11.3.1 Problem Description

Suppose an organization plans to promote a particular product (i_x) in an online social network. Due to limited resources, the organization needs to select a limited amount of influential users to experience the product and promote the product to the users connected with them. Ideally, the selected users can maximize the influence in the network. The purpose of IM is to select k initial vertices, also called seeds, in the social network. The prospective number of vertices influenced by the selected seeds is regarded as achieved influence.

11.3.2 Formal Definitions

In this chapter, we assume that there are m users, n edges, and x items in a social network. Here, a social network is modelled as a graph $G = (V, E)$, where V denotes the set of users, and E denotes the set of edges among the users. There are two types of edges in G, i.e., preference edge and trust edge. They represent UP and TC among users, respectively (refer to Definitions 5 and 6).

Definition 1 A **user** is defined as a vertex v_j in the network. Each user has a set of neighbours, i.e., $N_j = \{v_j \in V | v_j, v_i \in E\}$. Each vertex in N_j ($v_i \in N_j$) has a Trust Edge (refer to Definition 6) with v_j.

Beside users, the network also contains a set of items, i.e., I. An item $i_x \in I$ is a particular product which has been or will be promoted to the users in the network. User v_j's preference to item i_x is presented as the result of ratings given by v_j.

Definition 2 **Rating** r_{jx} is the preference degree of user v_j for item i_x. The rating set $R_j = (r_{j1}, r_{j2}, r_{j3}, \ldots, r_{jx})$ is the set of all ratings v_j once gave.

Definition 3 **Common Preference for Item** (cpi_{ijx}) is defined as ratings that any two users in the network gave for item they both rated, where cpi_{ijx} denotes item i_x is the common preference for user v_i and v_j.

$$cpi_{ijx} = 1 - \frac{|r_{ix} - r_{jx}|}{r_x.max - r_x.min} \tag{11.1}$$

In Eq. 11.1, r_{ix} is the rating to item i_x given by user v_i, and $|r_{ix} - r_{jx}|$ indicates the rating difference of i_x give by user v_i and v_j. While, $r_x.max - r_x.min$ is the gap of i_x's maximum and minimum rating value.

Definition 4 **Common Preference Similarity** (cps_{ij}) is defined as the similarity of ratings that users v_i and v_j in the network gave for all the item(s) they both rated. Users with common preference(s) will be computed for CPS and labelled CPS as weight on preference edges.

$$cps_{ij} = \frac{\sum_{x \in I} cpi_{ijx}}{I.count} \tag{11.2}$$

Equation 11.2 shows the CPS calculation between v_i and v_j. I denotes the item set rated by both v_i and v_j, I.count indicates the number of items in I.

Definition 5 A **Preference Edge** pe_{ij} denotes the preference relationship of two users, i.e. v_i and v_j. Users v_i and v_j will have a preference edge pe_{ij} when they have cpi_{ijx}. The weight of pe_{ij} can be denoted as $w(pe_{ij}) = cps_{ij}$.

Definition 6 A **Trust Edge** te_{ij} denotes the trust relationship of two users, i.e. v_i and v_j. The weight of te_{ij} can be represented as $w(te_{ij}) = user\ distance$, which is computed based on n-dimensional coordinate of information (attributes) in user profiles (n is depending on the number of the attributes). The range of user distance is from 0 to 1.

Definition 7 Influence Probability p_{ij} is defined as the likelihood of the occurrence of influence propagation from user v_i to user v_j. The range of p_{ij} is from 0 to 1. p_{ij} is computed by the product of cps_{ij} and user distance of users v_i and v_j, where presented as Eq. 11.3.

$$p_{ij} = w(pe_{ij}) \times w(te_{ij}) \tag{11.3}$$

Definition 8 A **Community** C_r, refers to a set of users in any scale that have common preference(s). Users in a community have a compact relationship related to common preference among each other, even for those who are not linked directly. A social network G can be partitioned into a number of communities, i.e., $G = C = C_1, C_2, C_3, \ldots, C_r$. Here we assume that there are no intersections between communities.

The reason why we conduct community partition is that, assuming there is a community C_i and we plan to propagate influence within C_i, selecting influential users from community C_i to propagate influence is more efficient than selecting influential users from other communities [5]. Each community can not only present the CPS between any two users in the network, but also the relationship related to common preference among a set of users in a community.

11.4 The Preference-Based Trust Independent Cascade (PTIC) Model

The framework of the PTIC Model is shown in Fig. 11.1. There are four modules in the PTIC Model, i.e., the Preference Learning Module (PLM), the Community Partition Module (CPM), the Trust Connectivity Computation Module (TCCM), and the Seed Selection Module (SSM). At the beginning of this model, CPSs (refer to Definition 4) between users will be computed and evaluated by the PLM. The computation of CPS can be regarded as the preparation of the CPM since CPS is the weight labelled on the preference edge. After the computation of CPS, the CPM will be conducted. Users will be partitioned into communities based on CPS. In addition, user distance will be computed based on user profiles as TC in the TCCM. Finally, the SSM will be conducted based on the output of CPM and the TCCM.

11.4.1 Preference Learning Module (PLM)

In this module, CPS is calculated based on the rating difference(s) of two users, e.g., v_i and v_j. The smaller the average of their rating difference(s) is, the higher the CPS will be. The $cps_i j$ between v_i and v_j can be calculated by using Eqs. 11.1 and 11.2.

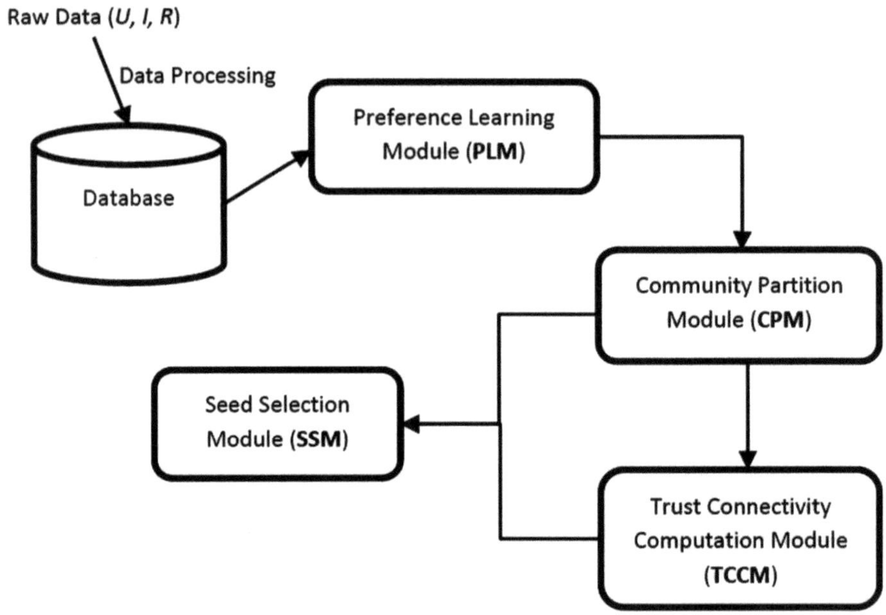

Fig. 11.1 Framework of the PTIC model

Algorithm 1 Common Preference Similarity (CPS) Computation Algorithm

Input: V, R
Output: CPS
1: Load the Users and Ratings from the process dataset
2: **for** $\forall v_i \in V$ **do**
3: **for** $\forall v_j \in V \wedge i > j$ **do**
4: Find common rating pairs $R', r_i, r_j \in R', R'.I_i = R'.I_j$
5: Calculate CPS between user v_i and user v_j, cps_{ij}
6: $cps_{ij} = cps_{ji}, cps_{ij} \in CPS, and\ cps_{ji} \in CPS$
7: **end for**
8: **end for**

Algorithm 1 shows the process for calculating the CPS. In the algorithm, the input includes user set V and user-item rating set R. We define network as an undirected network, therefore any user compares with the counterpart once only. The smaller the average of their rating difference(s) is, the higher their CPS will be.

As for the CPS, it represents the common preference similarity between any two users in the network (refer to Definition 4). As for the common preference similarity among the users inside a community, it can be obtained through community partition. Users with close CPSs will be partitioned into a community and each node in a community will have a relationship related to preference similarity among each other.In other word, community is used to represent common preference even if for those users who are not directly linked.

11.4.2 Community Partition Module (CPM)

In CPM, the community partition approach is derived from the community detection algorithm proposed by Ying et al. [8]. All of the users are randomized. Each individual tries to merge with the closest neighbour. This procedure will be conducted iteratively until similarity among the communities (refer to clusters in the algorithms) reaches a certain threshold σ.

Algorithm 2 User Preference Clustering Algorithm

Input: V, CPS
Output: C
Load the user dataset, V and user preference matrix, P
Randomize the user set
Initialize cluster, $C.size = V.size$
while average similarity among clusters $> \sigma$ **do**
    for $\forall v_i \in V$ **do**
        for $\forall v_j \in V \wedge i > j$ **do**
            Find v_j neighbour v_j with the maximum cps_{ij} in P
            if $cps_{ij} > threshold\ t_s$ **then**
                merge (v_i, v_j) into a new node v_n
                $V.size - 1$, assign v_n to a new cluster C_n
                for $\forall v_c \in \Gamma(v_i) \cap \Gamma(v_j)$ **do**
                    $cps_{nc} = max(cps_{cj}, cp_{ci})$
                end for
                for $\forall v_c \in \Gamma(v_j) \cup \Gamma(v_j)$ **do**
                    $cps_{nc} = cps_{ci} = cps_{cj}$
                end for
            end if
        end for
    end for
end while

The user preference clustering algorithm is shown in Algorithm 2. In this algorithm, there are two input variables, user set V and UP matrix P. While the output C indicates a tree-shape hierarchical UP cluster. The algorithm stops when the similarity among the communities reaches a certain threshold. For each round of iteration, nodes merge with their neighbours which have closest CPSs with those nodes. All the edges of merged nodes are updated accordingly. In the merging process, as for those common neighbours' edges, the one with higher weight is selected, while as for non-common neighbours', all the edges point to the merged node directly. After a number of iterations, a tree-structured cluster is generated.

11.4.3 Trust Connectivity Computation Module (TCCM)

The main purpose for computing TC is to ensure users with common preference(s) have influence propagation channel. Otherwise, even though users with common preference(s) are in the same community, if they do not have TC, influence cannot be propagated. TC is computed based on the information (attributes) in user profiles. Quantify the information when it is necessary and put all of them in n-dimensional coordinate (n is depending on the number of the attributes) to compute user distance as TC using Eq. 11.4.

$$w(te_{ij}) = 1 - \sqrt{\sum\nolimits_{a_m \in A} \left(\frac{v_i.a_m - v_j.a_m}{a_m.max - a_m.min}\right)^2} \tag{11.4}$$

In Eq. 11.4, $w(te_{ij})$ indicates the weight on the TC between user v_i and v_j, and $v_i.a_m - v_j.a_m$ denotes the attribute a_m distance between user v_i and v_j, where a_m is an element of the attribute set A.

Algorithm 3 User Trust Connectivity Computation Algorithm

Input: V, A
Output: T
Load the User dataset V including all the users' attribute set A
for $\forall v_i \in V$ **do**
    for $\forall v_j \in V \wedge i > j$ **do**
        Calculate the weight of TC between user v_i and v_j
        $w(te_{ij}) = w(te_{ji})$
    end for
end for

As mentioned above, the weight on TC is depending on the users' attributes. In Algorithm 3, the input is user set V and user attribute set A, while the output is user trust matrix T.

11.4.4 Seed Selection Module (SSM)

With the involvement of community partition and TC computation, users have not only common preferences, but also TC in the network. Influence probability p_{ij} is calculated based on the product of UP and TC between users v_i and v_j. After the computation of influence probability p_{ij}, seed selection is conducted and hub users who are interested in the promoted item will be selected as influential users (seed set). Seeds are selected based on heuristic methods in this chapter. Assume p number of seeds is going to be selected based on budget, the ones with high influence spread will be selected.

Algorithm 4 Influence Propagation Algorithm using IC Model

Input: $\{v_a\}, \{v_a\} \subseteq Seeds\ Set$
Output: V_a
1: Initialize $IPP = 1$ if not a recursive invoke
2: $v_a.activeStatus = true$
3: **for** $\forall v_i \in \Gamma(v_a)$ **do**
4:     **if** $v_a.activeStatus = true$ **then**
5:         Next
6:     **end if**
7:     **if** $p_{ai} \times IPP \geq propagation\ threshold$ **then**
8:         Generate a random decimal d_r, $0 \leq d \leq 1$
9:         **if** $d_r \leq p_{ai} \times IPP$ **then**
10:             $v_i.activeStatus = true$
11:             Update IPP, input v_i as variable and invoke self - Recursive
12:         **end if**
13:     **end if**
14: **end for**

Algorithm 4 aims to calculate the activated users influenced by the seed set in the network by using the IC model. The input is the seed set $\{v_a\}$, while the output is a set of activated users V_a in the entire network. p_{ai} denotes the influence probability between two users v_i and v_a (refer to Definition 7). In each iteration in Algorithm 4 (Lines 3–14), find the user v_a's neighbour set V_n, $v_i \in V_n$, if v_i is inactive and its Influence Propagation Probability (IPP) is larger than the threshold, then v_i is activated and the neighbours $\Gamma(v_i)$ are influenced with the IPP', $IPP' = IPP \times p_{ix}$ $(x \in \Gamma(v_i))$. The algorithm is recursive, and it invokes itself inside the method (Line 11).

For example, if v_a is an element of seed set, then $v_a's$ $IPP = 1$, the IPP of $v_a's$ neighbour v_i is $1 \times p_{ai}$, while the IPP of $v_i's$ neighbour v_j is $1 \times p_{ai} \times p_{ij}$. Hence IPP keeps reducing with the increment of hops of influence propagation.

11.5 Experiment and Analysis

In this chapter, two experiments have been conducted to compare the performance of the PTIC model with another two approaches, i.e., the random approach and the trust-only approach. In the random approach, the seeds are randomly selected from users. In the trust-only approach, seed selection is based on weight of TC only.

We estimated the total number of activated users influenced by the seed set that is generated by trust-only approach and random approach. If a user is selected into seed set, the user will intend to influence and activate the neighbours in a network. The activation probability undertaken by the neighbours is determined by the weight of the influence propagation channel.

In Experiment 1, the weight of influence propagation channel is determined by both UP and TC. Whereas, in Experiment 2, only TC is considered in influence propagation.

11.5.1 Data Selection

Movielens[1] dataset was used in the experiments, which is a stable benchmark dataset released on April of 1998. It contains the movie ratings of 1682 movies from 943 users. To filter noise data, users whose number of ratings are less than 50 have been removed from the dataset. Furthermore, users with ambiguous or false attributes are also eliminated from the experiments. After the data preprocessing, 441 users had been selected from the dataset.

11.5.2 Experimental Results

In both Figs. 11.2 and 11.3, x-axis denotes the size of seed set, i.e., the number of selected influential users, while y-axis refers to the number of activated users in the entire network.

Figure 11.2 shows the performance of three approaches in Experiment 1. There are 441 users, but only approximate 280 users are capable of activating the neighbours (the remaining users can only activate themselves). In order to ensure the accuracy of the individual's influential spread, we conduct the trials for 100 times and compute the average of it. As for the seed set with multiple elements, the same method has been utilised. The seed set increases by retaining the selected users and adding new users.

In Fig. 11.2, it can be seen that the PTIC model performs better than the other two approaches. By considering the cost performance, the appropriate size of the seed set is 10, the elbow point in the PTIC model. In addition, the number of activated users in the network reached 272, which is significantly better than the other two approaches.

Figure 11.3 presents the performance of three approaches in Experiment 2. It can be observed from the figure that the trust-only approach has the best performance. However, the PTIC Model still performs very well, and much better than the random approach. When the size of seed set is small (range from 4 to 6), the seeds selected via the trust-only approach are capable of activating approximate 300 users, while 250 users are activated in the PTIC Model. Furthermore, the amount of activated users of the PTIC Model is very close to the trust-only approach after the seed size reached 10.

[1] https://grouplens.org/datasets/movielens.

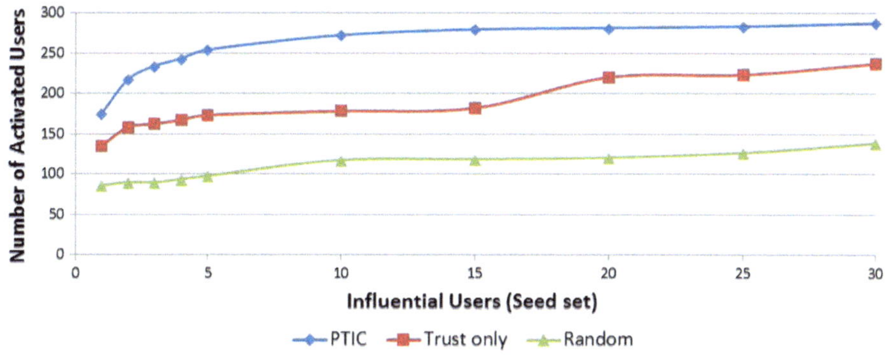

Fig. 11.2 Models comparison by considering user preference and trust

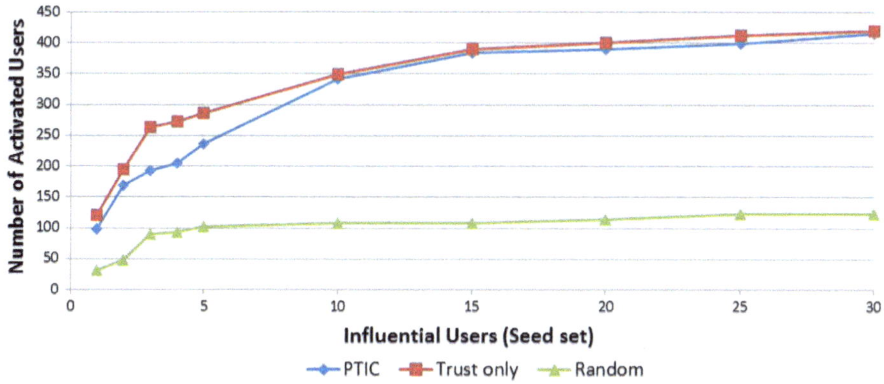

Fig. 11.3 Models comparison by considering trust only

Based on the above discussion, we could claim that when the resources are limited, the PTIC Model has a better performance.

11.6 Conclusion and Future Work

In this research, we intend to achieve IM on the foundation of UP and TC. In addition, compared with previous researches, we computed influence probability based on UP and TC rather than predefine it, which can improve the quality of seed selection. Based on the implementation of experiments, the results also present that the PTIC Model has a better performance. Hence, the PTIC model is able to propagate influence as much as possible based on UP and TC within limited resources efficiently. We implemented the PTIC Model based on IC Model. In the future, we are going to implement this model in other diffusion models, such as, LT Model [3]. In addition, we will consider trust relationship between users as directed.

References

1. Ahmed, S., Ezeife, C.I.: Discovering influential nodes from trust network. In: Proceedings of the 28th Annual ACM Symposium on Applied Computing, pp. 121–128. ACM, Coimbra, Portugal (2013)
2. Chen, W., Wang, Y., Yang, S.: Efficient influence maximization in social networks. In: Proceedings of the 15th ACM SIGKDD International Conference on Knowledge Discovery and Data Mining, pp. 199–208. ACM, New York, USA (2009)
3. Chen, W., Yuan, Y., Zhang, L.: Scalable influence maximization in social networks under the linear threshold model. In: Proceeding of the 10th IEEE International Conference on Data Mining (ICDM), pp. 88–97. IEEE, Washington, D.C., USA (2010)
4. Domingos, P., Richardson, M.: Mining the network value of customers. In: Proceedings of the 7th ACM SIGKDD International Conference on Knowledge Discovery and Data Mining, pp. 57–66. ACM, New York, USA (2001)
5. Fortunato, S.: Community detection in graphs. Phys. Rep. **486**(3), 75–174 (2010)
6. Huang, J., Sun, H., Han, J., Deng, H., Sun, Y., Liu, Y.: Shrink: a structural clustering algorithm for detecting hierarchical communities in networks. In: Proceedings of the 19th ACM International Conference on Information and Knowledge Management, pp. 219–228. ACM, New York, USA (2010)
7. Kempe, D., Kleinberg, J., Tardos, É.: Maximizing the spread of influence through a social network. In: Proceedings of the 9th ACM SIGKDD International Conference on Knowledge Discovery and Data Mining, pp. 137–146. ACM, New York, USA (2003)
8. Pan, Y., Li, D.-H., Liu, J.-G., Liang, J.-Z.: Detecting community structure in complex networks via node similarity. Phys. A: Stat. Mech. Appl. **389**(14), 2849–2857 (2010)
9. Phung, D.Q. Venkatesh, S., et al.: Preference networks: probabilistic models for recommendation systems. In: Proceedings of the 6th Australasian Conference on Data Mining and Analytics,-vol. 70, pp. 195–202. Australian Computer Society, Inc., Darlinghurst, Australia (2007)
10. Rashid, A.M., Albert, I., Cosley, D., Lam, S.K. McNee, S.M., Konstan, J.A., Riedl, J.: Getting to know you: learning new user preferences in recommender systems. In: Proceedings of the 7th International Conference on Intelligent User Interfaces, pp. 127–134. ACM, New York, USA (2002)
11. Richardson, M., Domingos, P.: Mining knowledge-sharing sites for viral marketing. In: Proceedings of the 8th ACM SIGKDD International Conference on Knowledge Discovery and Data Mining, pp. 61–70. ACM, New York, USA (2002)
12. Wang, C., Chen, W., Wang, Y.: Scalable influence maximization for independent cascade model in large-scale social networks. Data Min. Knowl. Discov. **25**(3), 545–576 (2012)
13. Zhou, J., Zhang, Y., Cheng, J.: Preference-based mining of top-k influential nodes in social networks. Future Gener. Comput. Syst. **31**, 40–47 (2014)

Chapter 12
Agent-Based Computation of Decomposition Games with Application in Software Requirements Decomposition

Jiamou Liu and Ziheng Wei

Abstract Coalition formation is a fundamental question in multiagent systems. The question asks for an optimal way in which agents may form coalitions and cooperate to accomplish a task. In this paper we investigate the use of coalition formation in software architecture design. We investigate a multiagent framework for attribute-driven software architecture design process. We design an agent for each requirement; the agents form coalitions that represent software components. The coalition formation process is based on decomposition game, a variant of coalition game. We extend previous work by adopting the propose-select-adjust framework for computing solutions of decomposition games. The focus is on analysing efficiency and utility of this agent-based approach. We also present three real-world case studies demonstrating the use of this approach to support software architecture design.

Keywords Coalition formation · Decomposition game · Software architecture · Software requirements · PSA-framework

12.1 Introduction

Coalition formation is the problem of *cooperation*: In a multiagent environment, the outcome of a task can often be improved if several agents join force to accomplish it together. The problem is important in a wide range of application domains such as task allocation [10] and electronic markets [13]. Therefore coalition formation has been a major topic of interest in multiagent systems [11]. *Coalition games* are useful tools to investigate coalition formation. Such games consist of a number of players, any non-empty subset of which is called a possible *coalition*. For each possible coalition, the game assigns a real number expressing its utility, which would then

J. Liu (✉) · Z. Wei
Department of Computer Science, The University of Auckland, Auckland, New Zealand
e-mail: jiamou.liu@auckland.ac.nz

Z. Wei
e-mail: zwei891@aucklanduni.ac.nz

Q. Bai et al. (eds.), *Multi-agent and Complex Systems*, Studies in Computational Intelligence 670, DOI 10.1007/978-981-10-2564-8_12

be used to derive payoffs of members of the coalition. The output of the game is a partition of all players, called a *coalition structure*, which should satisfy certain *stability* criterion, such as core, kernel, and Shapley value; these stability concepts all mean—to a degree—that the output coalition structure represents an equilibrium where the payoff of all agents are well-balanced and fair [1].

In this paper we investigate coalition formation as applied to the problem of *software architecture design*. Software architecture forms an important bridge between requirement analysis and more concrete software designs in the software engineering process; it defines a high-level software composition which meets function and quality requirements. The design and evaluation of software architectures typically demands high-level expertise and amounts to a largely manual task [3].

In our previous work [9], we initiated a game-based study of software architectures, hence providing a formal basis that supports the design and evaluation of software architectures. Our intuition is this: Software requirements exhibit *conflicts* and it is not possible to fulfill all requirements to a perfect degree; take, for example, security and performance, both are important quality feature of a software system. However, to ensure security, layers of encryption mechanisms should be put in place of a software system which harms performance. Therefore instead of finding the "perfect" software architecture, one would usually aim to find a "reasonable" software architecture that nicely balance all requirements. Abstractly, we may view software architecture design as a game, where players are requirements that may or may not be at odds of each other, and whose solutions are certain equilibrium states which satisfy all requirement to the best degree; such solution corresponds to a reasonable software architecture. Our approach blends two novel ideas:

(a) Firstly, we proposed a new game model, called a *decomposition game*, which follows the general setup of coalition games, but with the following important exceptions: Players are altruistic in the sense that they aim to maximise the utility of their coalitions, rather than their individual payoffs; the utility of a coalition depends solely on the interactions between its members which may either be beneficial or detrimental. The associated solution concept is called *rational decomposition*, and is computed using centralised algorithms.
(b) Secondly, we modeled the process of *attribute-driven design*—a method transforming software requirements to a conceptual software architecture—using decomposition games. Players of the game are software requirements and utility of a software component of is captured by the set of requirements it meets. The process of software architecture design is thus a process of coalition formation: Starting from the coalition of all requirements, we recursively divide the current collection into smaller coalitions, which are eventually mapped to meaningful software components.

The work [9] leaves some important questions unanswered. The first question concerns with the approach of finding a rational decomposition. A game entails that players are "acting on their own wills"; hence it is natural to ask if a multiagent approach (instead of a centralised approach) can be used to simulate the formation of coalitions. The second question concerns with the complexity of the computation.

The third questions concerns with how well the proposed approach may be adopted in practice. In this paper we aim to answer these questions:

1. We develop a multiagent environment for computing rational decompositions. Here we utilise the propose-select-adjust *framework* introduced in [7] and subsequently developed in [2, 8]. The approach involves two types of agents, coalitions and sub-coalitions, who autonomously decide on a coalition structure.
2. To achieve efficient computation, we pose a constraint T, which denotes the total amount of queries an agent is allowed to make within an iteration. This constraint helps to reduce the workload of a single agent in any iteration to constant time, and hence improves computation time. We demonstrate through experiments that under such constraints one can still identify reasonable decompositions.
3. We discuss the design of three real-world software systems: SplitPay system, Wargame2000 system and Cafeteria ordering system. For each case study we describe how our approach may help to identify a rational software architecture. The case studies demonstrate our method as a viable approach for deriving and evaluating software architectures, as well as trade-off analysis.

The rest of the paper is organised as follows: Sect. 12.2 presents decomposition games as the theoretical basis of our mechanism for coalition formation. Section 12.3 introduces the PSA-framework for realising coalition formation in a distributed approach, where agents starts from singleton coalitions and perform merge and bind operations to form larger coalitions. Section 12.4 discusses application of this coalition formation process in software architecture design. Section 12.5 presents experiments using synthetic games. Section 12.6 concludes the paper with a discussion on future works.

12.2 Decomposition Games

Let N be a set of players. A *coalition* is a non-empty subset of N. A *coalition structure* on N is a collection of coalitions $\mathscr{C} = \{C_1, C_2, \ldots, C_k\}$ such that $\bigcup_{i\geq 1}^{k} C_i = N$ and $C_i \cap C_j = \varnothing$ for any $i \neq j$. A *coalition game* models the situation when a group of agents form coalition based on individual payoffs; formally it is defined as a pair (N, v) where $v : 2^N \to \mathbb{R}$ is a *characteristic function* assigning every coalition to a utility. In the following, we consider a special type of coalition games, which have the following significant characteristics:

1. **Altruistic players**: We assume that players follow altruistic principle that values the collective utility of its coalition over individual payoffs. Therefore for any single player $a \in N$, the payoff of a equals to the utility of the coalition S that a belongs to. This is different from classical coalition games which assumes the sum of payoffs of members of a coalition S equals to the utility of S.
2. **Influence among players**: We assume that utility of a coalition is determined by the interactions among its members. For player a, there could be three types of

influence to another player b: (1) *a benefits b*, i.e., a has a positive effect on b; (2) *a detriments b*, i.e., a has a negative effect on b; (3) a is *independent* from b, i.e., a has no effect on b. Furthermore, we assume asymmetric relation so that the influence from a to b may be of different from the influence from b to a.

We define a decomposition game as follows. Let E be the set of ordered pairs of distinct players in N. Let $E_p \subseteq E$ be a set of *positive influence edges*. Let $E_n \subseteq E$ be a set of *negative influence edges*. We require that $E_p \cap E_n = \varnothing$. The *influence matrix* σ assigns every pair $(a, b) \in \mathbb{N}^2$ a value in $\{-1, 0, 1\}$ such that $\sigma(a, b) = 1$ if $(a, b) \in E_p$, $\sigma(a, b) = -1$ if $(a, b) \in E_n$ and $\sigma(a, b) = 0$ otherwise. A *relevance function* is $w : N \times N \to \mathbb{R}$ which measures the relevance $w(a, b)$ between two players a, b; we require $w(a, b) = w(b, a)$. The *interaction function* $\rho : N^2 \times 2^N \to \mathbb{R}$ denotes the level of interactions from player a to a player b within a coalition S, taking into account the influence from a to b:

$$\rho(a, b, S) = \begin{cases} \sigma(a, b) \times \sum_{a \neq c \in S} w(a, c), & \text{if}\, \sigma(a, b) \geq 0 \\ -\left| \sum_{a \neq c \in S} w(a, c) \right|, & \text{otherwise} \end{cases}$$

Intuitively, the sum $\sum_{a \neq c \in S} w(a, c)$ denotes the relevance of a to the coalition S. If a is independent from b, then the interaction from a to b is 0; if a benefits b, then this sum is the level of interaction from a to b; if a detriments b, then the negative value of this sum if the level of interaction.

Definition 1 (*Decomposition game*) A *decomposition structure* is $D = (N, E_p, E_n, w)$ as described above. A *decomposition game* of D is a coalition game $G = (N, v)$ where $v(S) = \sum_{a,b \in S} \rho(a, b, S)$.

In a decomposition game (N, v), players form a coalition structure based on their payoffs. We view the formation of coalitions as a dynamic process that starts with all players in singleton coalitions. In other words, if $N = \{a_1, \ldots, a_n\}$, the starting coalition structure is $\mathscr{C}_0 = \{\{a_1\}, \{a_2\}, \ldots, \{a_n\}\}$ where $N = \{a_1, \ldots, a_n\}$. The players then choose to form bigger coalitions if they can obtain higher payoff in this way. Suppose at some point the players arrive at a coalition structure $\{C_1, C_2, \ldots, C_k\}$. We assume that any individual player a's knowledge is bounded within her own coalition. This means, $a \in N$ only have knowledge of subsets $S \subseteq C_i$ where $a \in C_i$. Thus the players and coalitions may perform two strategies:

- *Merge*: Several coalitions may choose to merge if they would obtain a higher combined utility than their respective utilities.
- *Bind:* Several players within the same coalition may form a sub-coalition if they would obtain a higher utility than the current coalition.

The outcome of the decomposition game (N, v) is a coalition structure in which neither merge nor bind may take place:

Definition 2 (*Rational decomposition*) Let $\mathscr{C}$ be a coalition structure of N in decomposition game (N, v).

1. $\mathscr{C}$ is *merge-free* if for all $\mathscr{S} \subseteq \mathscr{C}$, $v\left(\bigcup \mathscr{S}\right) \leq \max\{v(C) \mid C \in \mathscr{S}\}$
2. $\mathscr{C}$ is *bind-free* if for all $C \in \mathscr{C}$, any $S \subseteq C$, $v(S) \leq v(C)$.

Call the coalition structure $\mathscr{C}$ a *rational decomposition* if it is both merge-free and bind-free.

A rational decomposition always exists and may not be unique. From a computation point of view, however, computing a rational decomposition is NP-hard [9]. Thus it makes sense to find ways to approximate a rational decomposition.

12.3 A Multiagent Framework for Decomposition Games

The propose-select-adjust (PSA) framework is a decentralized framework for simulating decision making among a network of agents [7, 8]. Generally speaking, each agent in this framework can be considered a utility-based agent, which maintains and updates its own state according its perception about the performance of possible future states. To carry out its computation, each agent repeatedly follows a simple three-step action of propose, select, and adjust:

- propose: The agent derives a *proposal* based on environment
- select: The agent collects proposals from other agents and selects a proposal with the highest performance
- adjust: The agent updates its own state according to the selected proposal

The above process is repeated by all agents in the network simultaneously and indefinitely, and thus an agent in PSA can also be regarded as a cell in a *graph dynamical system* [8]. At any time instance, the global state of the network (i.e., the collections of states of all agents) reveals the collective solution of all agents. In [2, 7], PSA have been applied to the problems of community detection in social networks, as well as market segmentation in consumer-commodity networks. It has been revealed that PSA is a viable framework for simulating and monitoring evolutions in dynamic networks. In this paper, we use PSA as the computation framework for obtaining coalition structures in decomposition games. Our PSA framework contains two types of agents:

- **Coalition agents**: At any time instance, the decomposition game (N, v) is at a particular coalition structure $\mathscr{C} = (C_1, \dots, C_k)$. Each coalition C_i is an agent which we call a *coalition agent*.
- **Sub-coalition agents**: For any coalition C_i, players in C_i also form a *sub-coalition structure* $\mathscr{D}_i = (D_1, \dots, D_\ell)$ of C_i. Each sub-coalition D_i is also an agent which we call a *sub-coalition agent*.

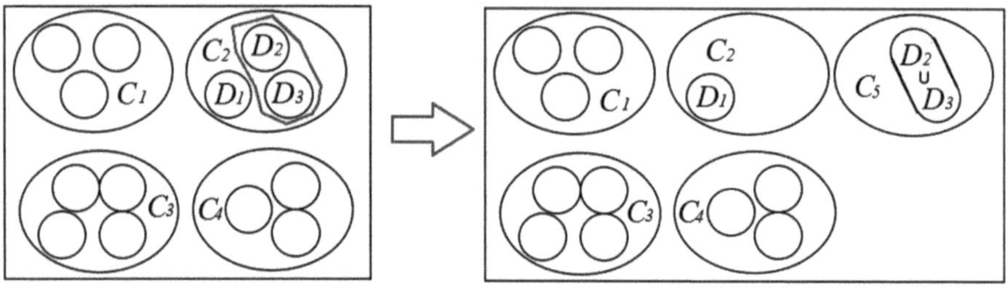

Fig. 12.1 Two sub-coalitions form a bigger sub-coalition whose utility exceeds their coalition, and thus breaks away to form a new coaltion

The two types of agents perform different propose, select, adjust steps.

- For a coalition agent C_i, its proposal $P(C_i)$ consists of a set of coalition agents $\{C_{j_1}, \ldots, C_{j_\ell}\}$ such that $v\left(C_i \cup \bigcup_{1\leq s\leq \ell} C_s\right) > v(C_i)$. A coalition agent would select the proposal with the highest utility made by other coalition agents that include itself. Once it selects a proposal $P(C_j)$, the coalition merges itself with the coalition agent C_j to form a larger coalition.
- For a sub-coalition agent D_i, its proposal $P(D_i)$ consists of a set of sub-coalition agents $\{D_{j_1}, \ldots, D_{j_\ell}\}$ such that each D_{j_s} and D_i belong to the same coalition, and $v\left(D_i \cup \bigcup_{1\leq s\leq \ell} D_s\right) > v(D_i)$. A sub-coalition agent would select the proposal with the highest utility made by other sub-coalition agents which include itself. Once it selects a proposal $P(D_j)$, the sub-coalition merges itself with the sub-coalition agent D_j to form a larger sub-coalition. As soon as any sub-coalition's utility exceeds the utility of the coalition that contains them, this sub-coalition "breaks away" from this coalition and lift itself into a coalition.

Figure 12.1 illustrates a possible scenario when two sub-coalitions bind to form a bigger sub-coalition and separate from their original coalition.

Let $\mathscr{C} = \{C_1, \ldots, C_k\}$ be a coalition structure. An *optimal proposal* of a coalition C_i is a set $P_o(C_i) = \{C_{j_1}, \ldots, C_{j_m}\}$ such that $v\left(C_i \cup \bigcup_{1\leq s\leq m} C_{j_s}\right)$ is maximal. For a sub-coalition structure $\mathscr{D}_i = \{D_1, \ldots, D_\ell\}$ of D_i, an *optimal proposal* $P_o(D_j)$ of a sub-coalition D_j can be defined in a similar way. The following theorem states that optimal proposals lead to the desired solution.

Theorem 1 *If the* PSA*-framework is implemented in such a way that all coalition and sub-coalition agents only make optimal proposals, then the players eventually stablise at a coalition structure $\mathscr{C}_o$. Furthermore, $\mathscr{C}_o$ is a rational decomposition.*

Searching for an optimal proposal for any agent may demand exhaustively looking through the space of all subsets of agents and thus is time-consuming. To ensure timely return of a proposal, we put a bound $T \in \mathbb{N}$ on the amount of information an agent is allowed to query in order to obtain a proposal. Under this constraint, a coalition agent C_i is not able to scan over arbitrary subsets of agents, but is restricted to examining only subsets of a bounded size α, defined as the largest value that satisfies $k^\alpha \leq T$, where $k = |\mathscr{C}|$, the number of coalitions in the current coalition structure $\mathscr{C}$. Thus we define the proposal $P_T(C_i)$ made by a coalition agent C_i as a subset of $\mathscr{C}$ with $|P_T(C_i)| \leq \alpha$ and

$$\forall P \subseteq \mathscr{C} : |P| \leq \alpha \Rightarrow \nu\left(C_i \cup \bigcup P_T(C_i)\right) \geq \nu\left(C_i \cup \bigcup P\right)$$

Similarly, a sub-coalition agent D within a sub-coalition structure $\mathscr{D}_i$ is not able to check arbitrary subsets of sub-coalitions in $\mathscr{D}_i$, but is restricted to examine subsets of a bounded size α, defined as the largest such that $\ell^\alpha \leq T$, where $\ell = |\mathscr{D}_i|$. We can then define the proposal $P_T(D)$ made by a sub-coalition agent D similarly to $P_T(C_i)$. The proposals $P_T(C_i)$ and $P_T(D)$ are referred to as *T-bounded optimal proposals.* The following is easy to check.

Proposition 1 *For a fixed $T \in \mathbb{N}$, computing a T-bounded optimal proposals for an agent takes constant time.*

Using T-bounded proposals to implement the PSA framework, the players would also reach a stablising coalition structure, which we call *T-bounded decompositions.* We will show using experiments in subsequent sections that T-bounded decompositions are very likely to be rational decompositions.

12.4 Game-Based Software Architecture Design

Decomposition game is introduced to formally capture the *attributed-driven design* (ADD) methodology. ADD amounts to a thorough and standardised paradigm for software architectures design after extensive development in the last 15 years [5, 15]. Inputs to ADD are software requirements and their relations, which are assumed to be derived from requirement analysis. There are two types of requirements: *Functional requirements*, which specify tasks the system performs, and *non-functional requirements*, which refer to quality attributes such as performance, security, availability, modifiability, usability, testability, and portability. We describe these quality attributes using *general scenarios*. The actual non-functional requirements are instances of general scenarios, simply called *scenarios*, which are real-world situations that refer to specific general scenarios. For example, "*If a failure occurs, the banking system notifies the user; the system continues to perform with half the efficiency*" is a scenario that refers to the availability requirement.

Requirements exhibit complicated relations. For example, "the system finalises a payment transaction" should be preceded by "correct user credentials are checked" (this is a form of dependency between functional requirements), and quality attributes such as ensuring security of the system typically harm its usability. Requirement analysis has identified influences between common quality attributes; see [14] for full description of the influence matrix σ. We note that the influence matrix is not symmetric: e.g., An improvement in performance may not affect security, but increasing security will almost always adversely impact performance.

An *attribute primitive* is a software component that meets several functional and non-functional requirements. Examples of attribute primitives include data router, firewall, virtual machine, interpreter and so on. In particular the entire software can be regarded as an attribute primitive that meets all requirements. The ADD methodology specifies a list of common attribute primitive along with their properties and side effects; see for example in [4]. We use F to denote the set of functional requirements and S to denote the set of scenarios (non-functional requirements). Let $\mathsf{R} = \mathsf{F} \cup \mathsf{S}$. A *design element* is a subset $C \subseteq \mathsf{R}$. A *decomposition* of an attribute primitive is a set of design elements $\{C_1, C_2, \ldots, C_k\}$ which form a coalition structure of all requirements met by the attribute primitive.

The ADD process can be viewed as a process of requirement decomposition: The process starts with the whole software as an attribute primitive, and derives a decomposition $\{C_1, C_2, \ldots, C_k\}$ of R. The process then assigns an attribute primitive $\mathscr{A}_i$ that meets each design element C_i. If $\mathscr{A}_i$ in turn requires further decomposition, the ADD process will then be recursively applied to $\mathscr{A}_i$. Therefore, the ADD process is the problem of deriving a suitable decomposition given the set of requirements R and their relations. We describe the ADD procedure as applied to an attribute primitive $\mathscr{A}$ recursively as follows:

Procedure 1 $\mathsf{ADD}(\mathscr{A})$ (General Plan)

```
1: (C_1, C_2, ..., C_k) ← Decompose(𝒜) // compute a rational decomposition of 𝒜
2: for 1 ≤ i ≤ k do
3:     𝒜_i ← an primitive attribute consistent with D_i
4:     if 𝒜_i needs further decomposition then
5:         ADD(𝒜_i)
```

To realise the $\mathsf{Decompose}(\mathscr{A})$ step in the ADD procedure, we apply our multiagent framework by letting each requirement $a \in \mathsf{R}$ act as a player and relevance $w(a, b)$ between to players a, b are determined by the relation between the corresponding requirements. A rational decomposition of the game is then the desired output of the process. For a thorough and more formal description of this agent-based model, the reader is referred to [9].

To demonstrate how our approach may be helpful to support software architecture design in real life, we provide three case studies below.

12.4.1 Case Study 1: SplitPay System

SplitPay is a mobile application based on Android platform. The purpose of this application is to manage shared expenses. Users in a group will post bills that they have paid for the group. Then, debts will be allocated to group users. The requirements of SplitPay are well documented in [12]. We elicit design elements from the documentation and organize them as three kinds of design driver: functional requirement,

Table 12.1 Coalitions in SplitPay system. The coalition C_3 has the highest utility which highlights the importance of usability

C_1	Performance1.1, Performance1.3, Function16.1, Function12.1, Function10.1
C_2	Function4.1, Function3.1, Function5.1, Performance1.2, Function15.1, Function14.1, Function11.1
C_3	Usability3.1, Function13.1, Function8.1, Function7.1, Function2.3, Function2.1, Function6.1, Function2.2, Function9.1, Function1.1
C_4	Performance1.4
C_5	Safety2.2

non-functional requirement and design constraints. Non-functional requirements in SplitPay's documentation are quite informal. We have to further analyze these non-functional requirements in order to correlate them with other design drivers. We use these elicited design drivers and their relationships as input to construct a decomposition game. Each design driver is a player in a decomposition game. Relationships between design drivers are weighted so that we are able to compose a evaluation function for every pair of design drivers. There are 18 functional requirements and 6 non-functional requirements. We compute a 10000-rational decomposition using PSA which converges in 6 iterations. The PSA-system has stablised on 5 coalitions. We show the results in Table 12.1. We only show the names of design drivers, where their exact meaning can be found in [12].

Based on this coalition structure, we derive a conceptual architecture as in Fig. 12.2. The decomposition highlights usability. User operations which require extra processing time will be handled in the daemon processes. Android systems can create service component for an application so that some processes will not affect user interface. C_2 can be a service component which listens from server and wraps up user requests to a formal HTTP request. C_3 is also a service component which updates a user's debt. C_1 is an application server which can install PHP and MySQL server.

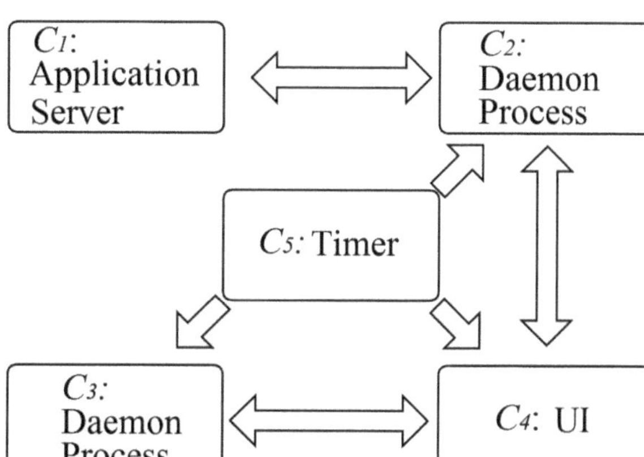

Fig. 12.2 SplitPay system architecture. *Rectangular boxes* represent system components and *arrows* represent their communications

Table 12.2 Coalitions in Wargame2000 system

C_1	Perfo3.1, Perfo3.2, Perfo3.3, Usabi4.1, Avala1.1, Avala1.2, Scala6.1, Scala6.2
C_2	Reliab2.1, Modif5.1, Modif5.2
C_3	Interop7.1

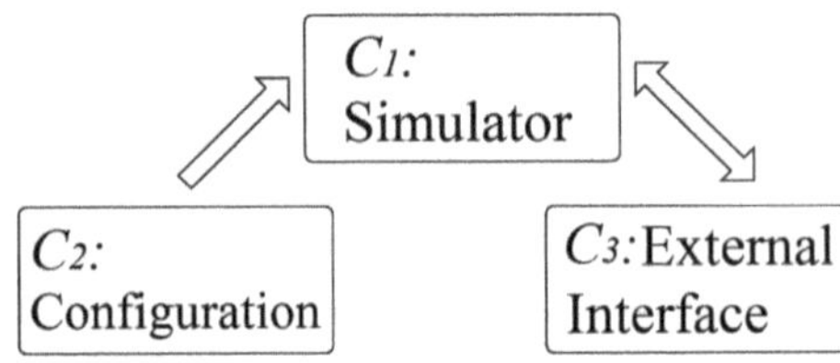

Fig. 12.3 Wargame system architecture

12.4.2 *Case Study 2: Wargame2000*

Wargame2000 is a highly complex real-time ballistic missile defense simulation system. Non-functional requirements of the systems are given in [6]; the focus in [6] is to apply architecture tradeoff analysis method (ATAM) to identify sensitive points and risks in the design phase. Therefore all functional requirements, design constraints are omitted except non-functional requirements, which are well described. A scenario describes a non-functional requirement by stating the stimulus, the environmental conditions, and the measurable or observable response to the stimulus.

Since we are only able to access non-functional requirements from [6], our focus here is also on tradeoff analysis. Our decomposition game contains 12 refined scenarios under 7 general scenarios that include availability, reliability, performance, usability, modifiability, scalability and interoperability. We establish interactions between these scenarios. For example, consider the two scenarios:

- *Availability*: Simulation controller initiates execution (a game), starts subroutine processes, loads parameter files, and simulation starts within 10 min.
- *Performance*: System must perform all initialization activities within 10 min.

These two scenarios are correlated because they both discuss the initialization procedure. We identify 48 such correlations between scenarios and compute a 10000-bounded decomposition, which is shown in Table 12.2.

Similar to [6], we identify a conceptual software architecture design in Fig. 12.3. This decomposition reveals a high level architecture of Wargame2000, which is presented in Fig. 12.2. Although this is a very coarse-grained architecture, it nevertheless reveals reasonable tradeoff among quality attributes. C_1 emphasises features for a simulator which requires real-time simulation and "human-in-loop" simulation. Some requirements (e.g. SCAL6.1) require a simulator to adapt new model. C_2 demands high quality of configuration of Wargame2000 system. C_3 builds a functionality to connect Wargame2000 to other systems.

Table 12.3 Coalitions in COS

C_1	SI1.1, SI1.2
C_2	ay.Method, Deliver.Location, Confirm.Response, Deliver.Select, Units.Multiple, Confirm.More
C_3	Menu.Available, Deliver.Times, Place.Register
C_4	USE1
C_5	AVL1, PER1, PER2, ROB1, SEC1, SEC2, USE2, SI1.3, SI2.1, SI2.2, SI2.4, CI1, CI2, UI2, UI3, SI2.3, SI2.5, Retrieve, Menu, Done.Patron, Deliver, Place, Done.Store, Pay, Done, Done.Cafeteria Done.Failure, Done.Inventory, Done.Menu, Confirm, Done.Time, Units
C_6	SEC4
C_7	Place.Date
C_8	Pay.Deliver, Pay.Deduct, Pay.Pickup
C_9	Menu.Date
C_{10}	SAF1
C_{11}	Confirm.Display
C_{12}	PER3, Place.Cutoff, Place.No, Deliver.Notimes, Units.TooMany, Pay.NG, Confirm. Prompt, Pay.OK

12.4.3 Case Study 3: Cafeteria Ordering System

The aim of this case study is to apply our game model to a business management system. Cafeteria ordering system is a widely used case study for software design (see [14] for detailed description). The project is well documented in the literature. It consists of 11 non-functional requirements which belong to 6 types: availability (AVL), performance (PER), security (SEC), Usability (USE), robustness (ROB), safety (SAF). There are 49 functional requirements and a number of design constraints. We set bounds for a bounded decomposition to 5×10^5. A PSA system solves the game as shown in Table 12.3.

A closer look reveals that it is reasonable to disregard some singleton coalitions. For example, USE1 in C_4 is from a business rule which only require the abidance of certain development standard; SAF1 in C_{10} is set up for indicating users' dietary precaution. These requirements are irrelevant to software architecture design. By removing these non-essential requirements, we have resolved a software architecture design in Fig. 12.4. This design is more detailed than the others. To work in an architectural level, requirements that we have elicited require more refinements. For example, *Coalition 11* is just singleton coalition and it can be combined into *Coalition 5*; *Coalition 9* and *coalition 7* also can be combined into *Coalition 12*. As we can see, decomposition game does not only provide architecture design but also help designer to identify problems in requirements.

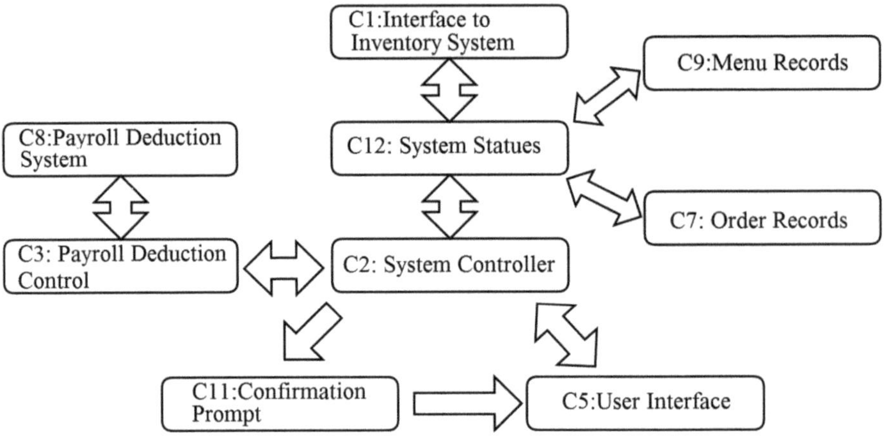

Fig. 12.4 COS architecture

12.5 Experiments

We conduct a number of experiments on synthetic games to evaluate the performance of our approach. Firstly, we calculate how much probability a bounded solution can be a core or a rational decomposition. Secondly, we measure the utilitarian welfare (sum of utilities) and the egalitarian welfare (the lowest utility) of the coalitions. Thirdly, we investigate the running times of our algorithm. We take the following parameters: a bound T for decompositions, the number n of players and a fixed value for α.

We fix a random model to generate decomposition games. The model firstly generates n players. Then, for each pair of players a and b, we randomly generate two weights within the range [0, 1]. One weight is for the direction from a to b and the other is for the opposite direction. For each weight, we take a 50 % probability to set it negative. We generate two types of games: *symmetric games* where the weights of the two directions (a, b) and (b, a) are the same, and *asymmetric games* where the two weights are not necessarily the same.

1. Results of Experiment 1 are shown in Fig. 12.5; here we investigate probability of a solution being in a core or a rational decomposition. In (a), we fix $n = 10$ players in the games while changing T from 200 to 2000. In (b), we also fix $n = 10$ players while changing α from 2 to 5 (note that this is different from our earlier description where α is determined by T, but rather is a fixed value). In (c), we fix $T = 2000$ while changing the number of players. The results show high probability of our approach finding a core or a rational decomposition. In addition, increasing T or α can both improve this probability. In particular, we notice in (c) that asymmetric games require much higher bounds for finding stable solutions if the number of players increases in decomposition games.

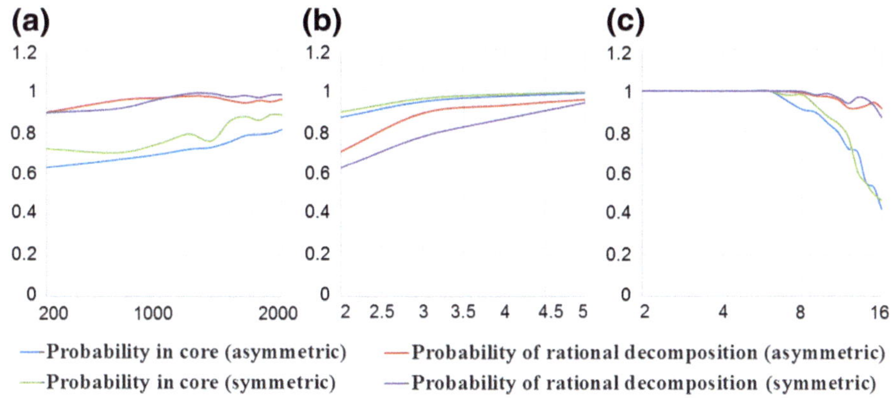

Fig. 12.5 Experiment 1: Comparison between probabilities of being in core and rational decomposition

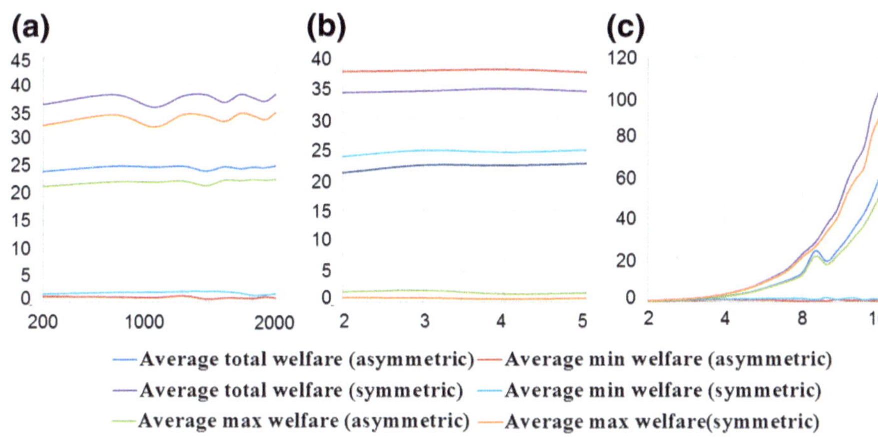

Fig. 12.6 Experiment 2: Achievements in utilitarian and egalitarian welfare

2. For Experiment 2, we use the same parameter setting as Experiment 1; the focus is to investigate utilitarian and egalitarian welfare obtained by decompositions games. The results of this experiment is shown in Fig. 12.6. The results in (a) and (b) change very little with varying parameters because a bounded solution has high probability in a core or a rational decomposition.
3. For Experiment 3, we follow the parameter setting from the previous two experiments; the focus here is to examine running times for finding a solution in decomposition games. The results of this experiment is shown in Fig. 12.7. With changing T in (a), the computation for each iteration is constance, and hence the time reflect the number of iterations before stability is reached. The running time shows a quadratic growth. With changing α in (b), the results are similar but

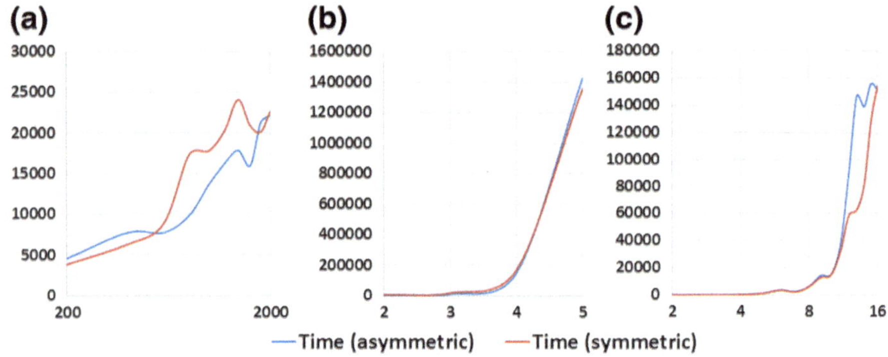

Fig. 12.7 Experiment 3: Time cost in changing T and α

running times increase exponentially. Thus we conclude that adopting the bound T to determine the values of α will lead to much faster computation while still achieving good accuracy.

12.6 Conclusion and Future Work

In this paper we simulate coalition formation based on decomposition games. The main difference between this work and our previous work [9] is that (1) here we use a multiagent framework that derive game solution in a distributed manner; (2) we put a bound on the complexity of query which improves efficiency. The case studies demonstrate that the approach can be used to support software architecture design by treating each requirement as a player. The experiments focus on performance of the solutions as well as time complexity and their results show that the multiagent framework can help us to achieve rational decompositions.

There are two directions for future works. Firstly, one may enrich the decomposition model by removing the assumption of altruistic agents and consider inhomogeneous individual payoffs. It would be interesting to investigate if common coalitional game solution concepts such as Shapley value give rise to meaningful software decompositions. Secondly, the current model assumes a waterfall model of software engineering process where the design process does not start until requirement analysis is finished. In other software engineering paradigms, requirements come and go in the software design phase and therefore we stipulate that the multiagent approach would help to dynamically determine software architectures.

References

1. Airiau, S.: Cooperative games and multiagent systems. Knowl. Eng. Rev. **28**(4), 381–424 (2013)
2. Bai, Q., Liu, J., Wei, Z.: Simulating and modeling dual market segmentation using PSA framework. In: Proceedings of 2nd International Workshop on Smart Simulation and Modelling for Complex Systems (2015)
3. Bass, L.: Software Architecture in Practice. Pearson Education India (2007)
4. Bass, L., Klein, M., Moreno, G.: Applicability of general scenarios to the architecture tradeoff analysis method (No. CMU/SEI-2001-TR-014). Carnegie-Melon University, Software Engineering Institute (2001)
5. Bass, L., Klein, M., Bachmann, F.: Quality attribute design primitives and the attribute driven design method. In: Proceedings of PFE-4, Revised Papers from the 4th International Workshop on Software Product-Family Engineering, pp. 169–186. Springer (2001)
6. Jones, L., Lattanze, A.: Using the architecture tradeoff analysis method to evaluate a wargame simulation system: A case study (No. CMU/SEI-2001-TN-022). Cargenie-Mellon University, Pittsburgh (2001)
7. Liu, J., Wei, Z.: From a local to a global perspective of community detection in networks. In: PRICAI 2014, Trends in Artificial Intelligence, pp. 1036–1049. Springer International Publishing (2014)
8. Liu, J., Wei, Z.: Community detection based on graph dynamical systems with asynchronous runs. In: 2014 Second International Symposium on Computing and Networking (CANDAR), pp. 463–469. IEEE (2014)
9. Liu, J, Wei, Z.: A game of attribute decomposition for software architecture design. In: Proceedings of the 12th International Colloquium on Theoretical Aspects of Computing (ICTAC 2015) (to appear) (2015). arXiv.org:1508.02812
10. Shehory, O., Kraus, S.: Methods for task allocation via agent coalition formation. Artif. Intell. **101**(1–2), 165–200 (1998)
11. Shoham, Y., Leyton-Brown, K.: Multiagent Systems: Algorithmic, Game-Theoretic, and Logical Foundations. Cambridge University Press (2008)
12. SplitPay SRS: https://www.cise.ufl.edu/class/cen3031sp13/SRS_Example_1_2011.pdf
13. Tsvetovat, M., Sycara, K., Chen, Y., Ying, J.: Customer coalitions in electronic markets, in: Agent-Mediated Electronic Commerce III, pp. 121–138. Springer (2001)
14. Wiegers, K., Beatty, J.: Software Requirements. Pearson Education (2013)
15. Wojcik, R., Bachmann, F., Bass, L., Clements, P., Merson, P., Nord, R., Wood, B.: Attribute-Driven Design (ADD), Version 2.0 (No. CMU/SEI-2006-TR-023). Carnegie-Melon University, Software Engineering Institute (2006)

Chapter 13
Preliminary Estimating Method of Opponent's Preferences Using Simple Weighted Functions for Multi-lateral Closed Multi-issue Negotiations

Shinji Kakimoto and Katsuhide Fujita

Abstract Multi-lateral multi-issue closed negotiation is an important class of real-life negotiations. Negotiation problems usually have constraints, such as an unknown opponent's utility in real time or time discounting. Recently, the attention in this field has shifted from bilateral to multi-lateral approaches. In multi-lateral negotiations, agents need to simultaneously estimate the utility functions of more than two agents. In this chapter, we propose an estimating method that uses simple weighted functions by counting the opponent's evaluation value for each issue. For multi-lateral negotiations, our agent considers some utility functions as the 'single' utility function by weighted-summing them. We experimentally compared the individual utility and the social welfare among some simple weighted functions. In addition, we compared the negotiation efficiency of our proposed agent with ten state-of-the-art negotiation agents that reached the final round of ANAC-2015.

Keywords Multilateral closed negotiations · Automated negotiation agent · Multi-issue negotiations

13.1 Introduction

Negotiation is a critical process in forming alliances and reaching trade agreements. Research in the field of negotiation originated in various disciplines including economics, social science, game theory, and artificial intelligence (e.g., [10, 16, 18] etc.). Automated agents, which can be used side-by-side with human negotiators who is embarking on an important negotiation task, can reduce some of the effort required by people during negotiations and assist those who are less qualified in the nego-

S. Kakimoto (✉) · K. Fujita
Faculty of Engineering, Tokyo University of Agriculture and Technology, Fuchu, Japan
e-mail: kakimoto@katfuji.lab.tuat.ac.jp

K. Fujita
e-mail: katfuji@cc.tuat.ac.jp

Q. Bai et al. (eds.), *Multi-agent and Complex Systems*, Studies in Computational Intelligence 670, DOI 10.1007/978-981-10-2564-8_13

tiation process. There may even be situations in which automated negotiators can replace the human negotiators [6, 22]. Another possibility is for people to use these agents as a training tool prior to actually performing the task. Thus, successfully developing an automated agent with negotiation capabilities has great advantages and implications for our field.

Motivated by the challenges of bilateral negotiations among automated agents, the automated negotiating agents competition (ANAC) was organized [1] to facilitate research in the area of automated multi-issue closed negotiation. ANAC's setup is a realistic model including time discounting, closed negotiations, alternative offering protocols, and so on. By analyzing the ANAC results, the trends of the strategies of automated negotiations and important factors for developing the competition have been shown [3]. Other effective automated negotiating agents have also been proposed through the competitions [5, 8].

Having multiple parties is a key point in achieving automated negotiation in real life. Many real-world negotiation problems assume multi-party situations because negotiations on the web are becoming more common. When an automated negotiation strategy effectively covers the bilateral negotiation, it is not always possible or desirable in multi-party negotiations. In other words, it remains an open and interesting problem to design more efficient automated negotiation strategies against a variety of negotiating opponents in multi-party situations.

In this chapter, the negotiation protocol adopts a simple extension of the bilateral alternating offers protocol that is called the Stacked Alternating Offers Protocol for Multi-lateral Negotiation (SAOP) [2]. According to it, all of the participants around the table get a turn per round; turns are taken clock-wise around the table. In addition, we propose an estimating method using simple weighted functions by counting the opponent's evaluation value of each issue. For multi-lateral negotiations, our agent considers some utility functions as the 'single' utility function by weighted-summing them. In our experiments, we compare the individual utility and the social welfare among some simple weighted functions and the negotiation efficiency of our proposed agent with ten state-of-the-art negotiation agents that reached the final round of ANAC-2015.

The remainder of the chapter is organized as follows. First, we describe related works and show the negotiation environments and the Stacked Alternating Offers Protocol for Multi-lateral Negotiation (SAOP). Next we propose a novel method for estimating opponent utility functions by weighted-summing the value of each issue. Then we demonstrate our experimental analysis and the tournament results. Finally, we present our conclusions.

13.2 Related Works

This chapter focuses on research in the area of bilateral multi-issue closed negotiation, which is an important class of real-life negotiations. In closed negotiations, opponents do not reveal their preferences. Negotiating agents that are designed using a heuristic approach require extensive evaluation, typically through simulations and empirical analysis, since it is usually impossible to predict precisely how the system and its constituent agents will behave in a wide variety of circumstances. Motivated by the challenges of bilateral negotiations between people and automated agents, the automated negotiating agents competition (ANAC) was organized in 2010 [1] to facilitate research in the area of bilateral multi-issue closed negotiation.

The following are the declared goals of the competition: (1) to encourage the design of practical negotiation agents that can proficiently negotiate against unknown opponents in a variety of circumstances; (2) to provide a benchmark for objectively evaluating different negotiation strategies; (3) to explore different learning and adaptation strategies and opponent models; (4) to collect state-of-the-art negotiating agents and negotiation scenarios and make them available to the wider research community. The competition is based on the GENIUS environment: the General Environment for Negotiation with Intelligent multi-purpose Usage Simulation [17]. By analyzing the ANAC results, the trends of ANAC strategies and important factors for developing the competition have been shown. Baarslag et al. presented an in-depth analysis and the key insights gained from ANAC 2011 [3] and analyzed different strategies using the classifications of agents with respect to their concession behavior against a set of standard benchmark strategies and empirical game theory (EGT) to investigate the robustness of the strategies. Even though most adaptive negotiation strategies are robust across different opponents, they are not necessarily the ones that win the competition. Our EGT analysis highlights the importance of considering metrics.

Chen and Weiss proposed a negotiation approach called OMAC, which learns an opponent's strategy to predict the future utilities of counter-offers by discrete wavelet decomposition and cubic smoothing splines [7]. They also presented a negotiation strategy called EMAR for such environments that rely on a combination of Empirical Mode Decomposition (EMD) and Autoregressive Moving Average (ARMA) [8]. EMAR enables a negotiating agent to acquire an opponent model and to use it for adjusting its target utility in real time on the basis of an adaptive concession-making mechanism. Hao and Leung proposed and introduced a negotiation strategy named ABiNeS for negotiations in complex environments [12]. ABiNeS adjusts the time to stop exploiting negotiating partners and also employs a reinforcement-learning approach to improve the acceptance probability of its proposals. Williams et al. proposed a novel negotiating agent based on Gaussian processes in multi-issue automated negotiation against unknown opponents [21]. Baarslag et al. focused on the acceptance dilemma; accepting the current offer may be suboptimal, since better offers might still be presented [4]. Kawaguchi et al. proposed a strategy for compromising on the estimated maximum value based on estimated maximum utility [14]. These papers made important contributions to bilateral multi-issue closed negotiation; however, they failed to deal with multi-times negotiation. Fujita [11] proposed a compromising strategy by adjusting the speed of reaching agreements using the Conflict Mode and focused on multi-time negotiations.

13.3 Negotiation Environments

13.3.1 *Multi-lateral Multi-issue Closed Negotiation*

The interaction among negotiating parties is regulated by a *negotiation protocol* that defines the rules of how and when proposals can be exchanged. The competition used the alternating offers protocol for bilateral negotiation [19, 20] in which the negotiating parties exchange offers in turns. The alternating offers protocol, which conforms to our criterion that advocates simple rules, has been widely studied in the literature, both in game-theoretic and heuristic settings of negotiation [9, 10, 15, 18].

The multi-player protocol is a simple extension of the bilateral alternating offers protocol, called the Stacked Alternating Offers Protocol for Multi-lateral Negotiation (SAOP) [2]. According to this protocol, all of the participants around the table get a turn per round; turns are taken clock-wise around the table.

We assume that N agents ($A_1, A_2, \ldots, A_N$) negotiate at the same time. First A_1 starts the negotiation with an offer that is immediately observed by all the others. Whenever an offer is made, A_n in line can take the following actions:

- Make a Counter-offer (rejecting and overriding the previous offer)
- Accept the Offer
- Walk Away (ending the negotiation without any agreement)

After that, $A_{mod(n,N)+1}$ selects its next action from *Counter-offer, Accept, Walk Away*. This process is repeated in clock-wise turns until an agreement is reached or the deadline passes. To reach an agreement, all parties must accept the offer. If at the deadline no agreement has been reached, the negotiation fails.

Figure 13.1 shows an example of SAOP when the number of agents is three: (1) agent 1 makes a counter-offer; (2) agent 2 accepts agent 1's offer; and (3) agent 3

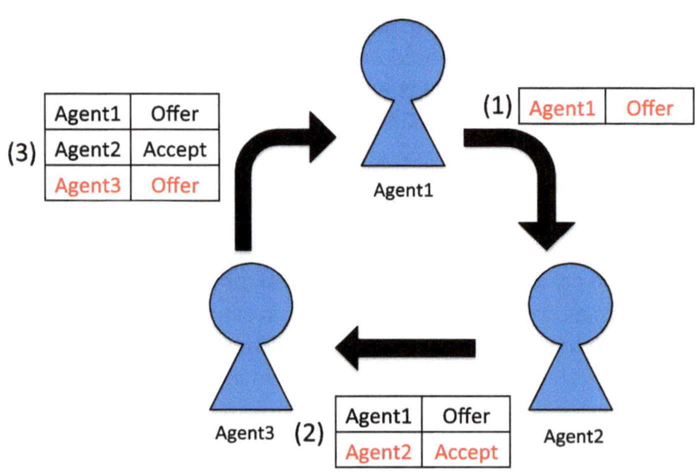

Fig. 13.1 Example of SAOP

makes a counter-offer by rejecting the last offer by agent 1. In this negotiation, the agreement failed because agent 3 selects a new counter-offer.

The parties negotiate over *issues*, each of which has an associated range of alternatives or *values*. A negotiation outcome consists of a mapping of every issue to a value, and set Ω of all possible outcomes is called the negotiation *domain*. This domain is the common knowledge shared by the negotiating parties and remains fixed during a single negotiation session. All parties have certain preferences prescribed by a *preference profile* over Ω. These preferences can be modeled by utility function U that maps possible outcome $\omega \in \Omega$ to a real-valued number in range $[0, 1]$. In contrast to the domain, the preference profile of the players is private information.

A negotiation lasts a predefined time in seconds (*deadline*). The time line is normalized, i.e., time $t \in [0, 1]$, where $t = 0$ represents the negotiation's start and $t = 1$ represents the deadline. Apart from a deadline, a scenario may also feature discount factors that decrease the utility of the bids under negotiation as time passes. Let d in $[0, 1]$ be the discount factor. Let t in $[0, 1]$ be the current normalized time, as defined by the timeline. We compute discounted utility U_D^t of outcome ω from undiscounted utility function U as follows:

$$U_D^t(\omega) = U(\omega) \cdot d^t.$$

At $t = 1$, the original utility is multiplied by the discount factor. If $d = 1$, the utility is not affected by time, and such a scenario is considered to be undiscounted.

13.3.2 Weighted-Summing Linear Utility Function

A bid is a set of chosen values $s_1 \dots s_N$ for each of the N issues (I). Each of these values has been assigned evaluation value $eval(s_i)$ in the utility space, and each issue has also been assigned a normalized weight (w_i, $\sum_{i \in I} w_i = 1$) in the utility space. The utility is the weighted-sum of the normalized evaluation values (Fig. 13.2).

The utility function of the bid ($\mathbf{s} = (s_1, \dots, s_N)$) is defined as Eq. (13.1):

$$U(\mathbf{v}) = \sum_{i=1}^{N} w_i \cdot eval(s_i).$$

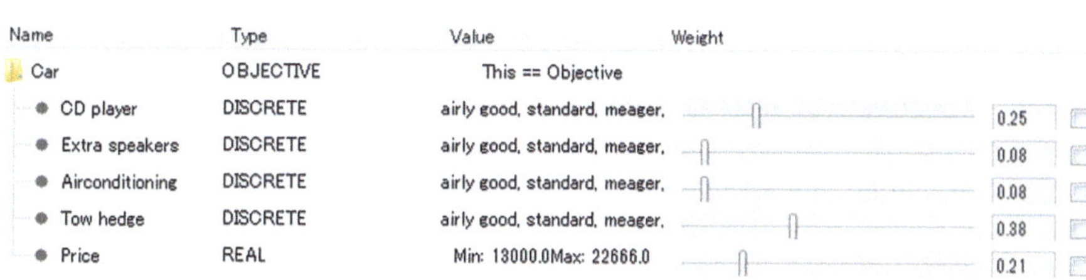

Fig. 13.2 Example of utility profiles

A negotiation lasts a predefined time in seconds (*deadline*). The timeline is normalized, i.e., time $t \in [0, 1]$, where $t = 0$ represents the negotiation's start and $t = 1$ represents the deadline. Apart from a deadline, a scenario may also feature discount factors that decrease the utility of the bids under negotiation as time passes. Let d in $[0, 1]$ be the discount factor. Let t in $[0, 1]$ be the current normalized time, as defined by the timeline. We compute discounted utility U_D^t of outcome ω from undiscounted utility function U as follows:

$$U_D^t(\omega) = U(\omega) \cdot d^t.$$

At $t = 1$, the original utility is multiplied by the discount factor. If $d = 1$, the utility is not affected by time, and such a scenario is considered to be undiscounted.

13.3.3 *Objective Functions of Negotiating Agents*

Objective functions of negotiating agents are as follows:

- Objective Function 1 (Maximizing Individual Welfare):

$$\max_{\mathbf{s}} \ U_a(\mathbf{s})$$

- Objective Function 2 (Maximizing Social Welfare):

$$\max_{\mathbf{s}} \sum_{i=1}^{N} U_{A_i}(\mathbf{s})$$

The agent, in other words, tries to find contracts that maximize the social welfare, i.e., the total utilities for all agents. Such contracts, by definition, will also be Pareto-optimal. At the same time, each agent tries to find contracts where individual welfare exceeds the reservation value.

13.4 Negotiation Strategy with Estimating Utility Functions by Counting Values

13.4.1 *Estimating Utility Functions by Counting Values*

In SAOP, the bids repeatedly proposed by opponents are critical. However, it is hard to get statistical information by simply counting all of them because proposed bids are limited in one-shot negotiations. Therefore, we propose a novel method that estimates the utility functions by counting the value of the opponent's bids in multi-lateral negotiations.

In our definitions, A_N is our agent and $a(a = \{A_1, A_2, \ldots, A_{N-1}\})$ are the $N-1$ opponents among the N-lateral negotiations. Agent a's previous bids are represented as B_a. The estimated utility function of agent a is represented as $eval'_a()$, which is defined as Eq. 13.1:

$$eval'_a(\mathbf{s_i}) = \sum_{\mathbf{s}' \in B_a} Boolean(\mathbf{s}_i, \mathbf{s}') \cdot \mathbf{w}(\mathbf{s}'). \tag{13.1}$$

The function $Boolean(\mathbf{s}_i, \mathbf{s}')$ returns 1 when bid s' contains the s_i, and otherwise it returns 0. Function $w(\mathbf{s})$ is the weighting function that reflects the order of the proposed bids. Therefore, estimated utility function $U'_a(\mathbf{s})$ of Alternative solutions: $\mathbf{s}$ of opponent a is defined as Eq. 13.2:

$$U_a(\mathbf{s}) = \frac{u_a(\mathbf{s})}{\max_{\forall \mathbf{s}'} u_a(s')} \tag{13.2}$$

$$u_a(\mathbf{s}) = \sum_{i=1}^{N} eval'_a(s_i). \tag{13.3}$$

Using Eq. 13.2, our agent can obtain estimated utility that is normalized [0, 1] to each opponent.

In addition, we use the following three weighting functions:

- Constant Function: $w(\mathbf{s}) = 1$
- Monotonically Increasing Function: $w(\mathbf{s}) = time_a(\mathbf{s})$
- Monotonically Decreasing Function: $w(\mathbf{s}) = 1 - time_a(\mathbf{s})$.

$time_a(\mathbf{s})$ is a function that returns the normalized time when agent a proposes bid $\mathbf{s}$.

13.4.2 Strategy of Negotiating Agents Using Estimated Utility Functions

We propose an agent's strategy using the estimated utility function in Sect. 13.4.1. In multi-lateral negotiations, a novel strategy needs to determine how much our agent can compromise to each agent using the estimated utility functions. Our proposed agent employs $h_n(n = 1, 2, \ldots, N)$, which is a function for compromising to each agent$(A_1, A_2, \ldots, A_N)$, to judge subsequent bids. The evaluation function (U_{op}) of the bid $(\mathbf{s})$ that combines two opponents is defined as Eq. 13.4:

$$U_{op}(\mathbf{s}) = \sum_{n=1}^{N} h_n U_{A_n}(\mathbf{s}). \tag{13.4}$$

In this chapter, the proposed method focuses on negotiations among three agents same as the ANAC2015 rule. However, our proposed method can adopt the multi-lateral negotiations not only the negotiations among three agents. Our agent can adopt a negotiation strategy for bilateral negotiations to multi-lateral negotiations by combining the utility functions of opponents. Our agent decides its next action based on Eqs. 13.5 and 13.6:

$$target_{end} = \frac{\sum_{n=1}^{N} h_n U_{my}(arg\ max U_{A_n})}{\sum_{n=1}^{N} h_n} \tag{13.5}$$

$$target(t) = \begin{cases} (1-t^3)(1-target_{end}) + target_{end} & (d=1) \\ (1-t^d)(1-target_{end}) + target_{end} & (otherwise) \end{cases} \tag{13.6}$$

The weighting average of the estimated value proposed by the opponent in the final phase divided by h_n is calculated using Eq. 13.5. Our agent proposes a bid whose utility exceeds $target(t)$ and the highest U_{op} (Eq. 13.4). It accepts the opponent's bids when they are more than $target(t)$.

Figure 13.3 shows the changes of $target(t)$ when $target_{end} = 0.5$. Since discount factor d is small, our agent compromises soon by considering the conflicts among agents based on $target_{end}$.

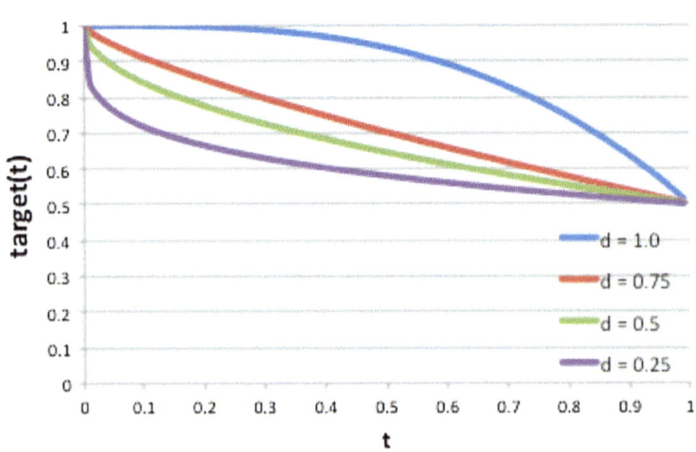

Fig. 13.3 Example of threshold of offer and accept ($target_{end} = 0.5$)

13.5 Experimental Results

Our experiments are demonstrated under the domains generated randomly. The number of values for each issue is 10. The number of issues is from 2 to 7 (Domain size is from 10^2 to 10^7). In each domain, the discount factor is set to 1.0 (without discounting) or random value (from 0.01 to 0.99), respectively. The reservation values is set to 0.

The opponents are the top five state-of-the-art agents in the individual utility and social welfare categories in ANAC2015.

- Individual Utility Category: agentBuyogV2 (Nanyang Technological University), Atlas3 (Nagoya Institute of Technology), kawaii (Nagoya Institute of Technology), ParsAgent (University of Isfahan), RandomDance (Tokyo University of Agriculture and Technology)
- Social Welfare Category: Mercury (Maastricht University), AgentX (Nagoya Institute of Technology), Atlas3 (Nagoya Institute of Technology), JonnyBlack (University of Tulsa), CUHKAgent2015 (The Chinese University of Hong Kong)

In these experiments, we compared our proposed agent with a random agent (*Random*) that doesn't estimate the opponent's utility. *Random* proposes bids over the threshold based on Eq. 13.5 whose maximum utility is the initial opponent's bid using the hill climbing algorithm. The tournament has three agents. The averages and standard deviations of the four tournaments under the ANAC2015 agents and domains are shown.

Figure 13.4 shows the individual utility (left graph) and the social welfare (right graph) using three types of $w(\mathbf{s})$ as the number of issues changes. The individual utility rate and the social welfare rate are defined as (*Utility of Three Types of* $w(\mathbf{s})$)/(Utility of Random Agent). In left graph of Fig. 13.4, the individual utility rate is higher as the domain size becomes large. Therefore, the method of estimating opponent's utility function is effective in the large sized domains. On the other hands, the social welfare rate is almost same as the domain size changes. This is because that the agents by finalists estimate the opponent's utility functions to get the higher social welfare.

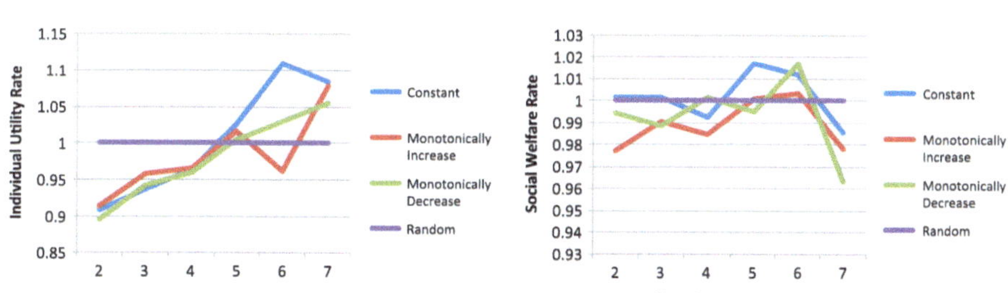

Fig. 13.4 Individual utility (*left*) and social welfare (*right*) using different $w(\mathbf{s})$

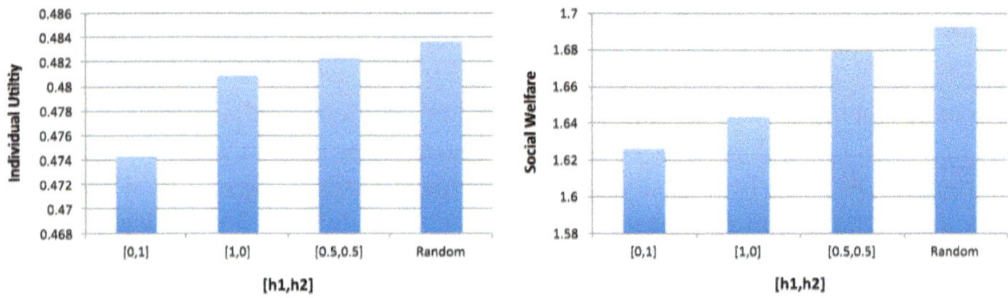

Fig. 13.5 Individual utility (*left*) and social welfare (*right*) using different h_n

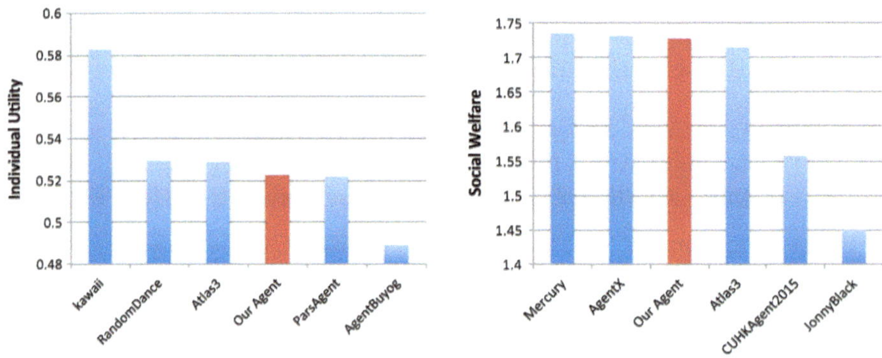

Fig. 13.6 Individual utility (*left graph*) and social welfare (*right graph*) under tournaments among ANAC2015 Agents

Figure 13.5 compares the result when the compromising function to each agent is changed. The following are the h_1 and h_2 rates: $(h_1, h_2) = (0.5, 0.5), (1, 0), (0, 1), (Random)$. The weighted function $w(\mathbf{s})$ is constant function. In individual utility and social welfare, $(h_1, h_2) = (Random)$ is the highest among all of the methods. In addition, $(h_1, h_2) = (1, 0), (0, 1)$ is lower than (0.5, 0.5) because the proposed agent compromises too much without considering another opponent's agent.

Figure 13.6 shows the individual utility (left graph) and the social welfare (right graph) under tournaments among the ANAC2015 agents. The weighting function is constant and the rate of the compromising function to each agent $(h_1, h_2) = (Random)$. Table 13.1 shows the individual utility (left table) and the social welfare (right table) only the largest domain in this experiments (Number of issues is 7, Domain size is 10^7). A rank of individual utility in the seven issues s a higher than the average in all domains. In large domain, our agent is effective despite that our agent uses the simple estimating method of opponent's utility function. However, these results aren't so high in this tournament because our proposed agent has a simple compromising strategy and constant weighting functions. In future work, we will improve our agent to get higher social welfare and individual utility.

Table 13.1 Individual utility (left table) and social welfare (right table) in issues size = 7 (Domain size = 10^7)

Agent Name	Individual Utility
kawaii	0.2047
Our Agent	0.1961
RandomDance	0.1766
Atlas3	0.1636
ParsAgent	0^a
AgentBuyog	0^a

Agent Name	Social Welfare
AgentX	0.6716
Atlas3	0.6704
Our Agent	0.6680
Mercury	0.6621
CUHKAgent2015	0^a
JonnyBlack	0^a

[a]These agents didn't work well in the large domains because the domains size of ANAC2015 is less than 2^{16}

13.6 Conclusions

This chapter focused on both multi-lateral multi-issue closed negotiations and bilateral negotiations. In multi-lateral negotiations, agents simultaneously estimated the utility functions of more than two agents. In this chapter, we proposed an estimating method using simple weighted functions by counting the opponent's evaluation value of each issue. For multi-lateral negotiations, our agent considered utility functions as a single utility function by weighted-summing them. In our experiments, we compared the individual utility and the social welfare among some simple weighted functions. We also evaluated the negotiation efficiency of our proposed agent with ten state-of-the-art negotiation agents that reached the final round of ANAC-2015.

Future works will improve the estimates of the opponent's utility in our proposed approach. To solve this problem, our approach needs to adjust the weighting functions based on opponent proposals for estimating the opponent's utility. In the experiments, we focused on the linear utility functions same as ANAC2015 rules. In the bilateral multi-issue closed negotiations, the authors proposed the method of estimating the Pareto fronts with nonlinear domains [13]. By considering these results, our approach can be improved to the nonlinear utility functions. Another important task is to judge the opponent's strategy based on modeling or a machine learning technique to further enhance our proposed method.

References

1. Aydogan, R., Baarslag, T., Fujita, K., Hindriks, K., Ito, T., Jonker, C.: The Fifth International Automated Negotiating Agents Competition (ANAC2015). http://www.tuat.ac.jp/~katfuji/ANAC2015/ (2015)
2. Aydoğan, R., Festen, D., Hindriks, K.V., Jonker, C.M.: Alternating offers protocols for multilateral negotiation. In: Fujita, K., Bai, Q., Ito, T., Zhang, M., Hadfi, R., Ren, F., Aydoğan, R. (eds.) Modern Approaches to Agent-based Complex Automated Negotiation. Springer (2016) (to appear)

3. Baarslag, T., Fujita, K., Gerding, E., Hindriks, K., Ito, T., Jennings, N.R., Jonker, C., Kraus, S., Lin, R., Robu, V., Williams, C.: Evaluating practical negotiating agents: results and analysis of the 2011 international competition. Artif. Intell. J. (AIJ) **198**, 73–103 (2013)
4. Baarslag, T., Hindriks, K.: Accepting optimally in automated negotiation with incomplete information. In: Proceedings of the 2013 International Conference on Autonomous Agents and Multi-agent Systems (AAMAS2013), pp. 715–722 (2013)
5. Baarslag, T., Hindriks, K.V.: Accepting optimally in automated negotiation with incomplete information. In: Proceedings of the 2013 International Conference on Autonomous Agents and Multi-agent Systems, pp. 715–722. International Foundation for Autonomous Agents and Multiagent Systems (2013)
6. Chavez, A., Maes, P.: Kasbah: an agent marketplace for buying and selling goods. In: Proceedings of the First International Conference on the Practical Application of Intelligent Agents and Multi-Agent Technology, vol. 31, p. 40. London, UK (1996)
7. Chen, S., Weiss, G.: An efficient and adaptive approach to negotiation in complex environments. In: Proceedings of the 19th European Conference on Artificial Intelligence (ECAI-2012), vol. 242, pp. 228–233 (2012)
8. Chen, S., Weiss, G.: An efficient automated negotiation strategy for complex environments. Eng. Appl. Artif. Intell. (2013)
9. Faratin, P., Sierra, C., Jennings, N.R.: Using similarity criteria to make issue trade-offs in automated negotiations. Artif. Intell. **142**, 205–237 (2002)
10. Fatima, S.S., Wooldridge, M., Jennings, N.R.: Multi-issue negotiation under time constraints. In: Proceedings of the First International Joint Conference on Autonomous Agents and Multi-agent Systems (AAMAS 2002), pp. 143–150. New York, NY, USA (2002)
11. Fujita, K.: Automated strategy adaptation for multi-times bilateral closed negotiations. In: Proceedings of the 13th International Conference on Autonomous Agents and Multiagent Systems (AAMAS 2014), pp. 1509–1510 (2014)
12. Hao, J., Leung, H.-F.: Abines: an adaptive bilateral negotiating strategy over multiple items. In: 2012 IEEE/WIC/ACM International Conferences on Intelligent Agent Technology (IAT-2012), vol. 2, pp. 95–102 (2012)
13. Kakimoto, S., Fujita, K.: Estimating pareto fronts using interdependency between issues for bilateral multi-issue closed nonlinear negotiations. IEICE Trans. Inf. Syst. (Japan. Ed.) **J98-D**(6), 926–935 (2015)
14. Kawaguchi, S., Fujita, K., Ito, T.: Compromising strategy based on estimated maximum utility for automated negotiation agents competition (ANAC-10). In: 24th International Conference on Industrial Engineering and Other Applications of Applied Intelligent Systems (IEA/AIE-2011), pp. 501–510 (2011)
15. Kraus, S.: Strategic Negotiation in Multiagent Environments. MIT Press (2001)
16. Kraus, S., Wilkenfeld, J., Zlotkin, G.: Multiagent negotiation under time constraints. Artif. Intell. **75**(2), 297–345 (1995)
17. Lin, R., Kraus, S., Baarslag, T., Tykhonov, D., Hindriks, K., Jonker, C.M.: Genius: an integrated environment for supporting the design of generic automated negotiators. Comput. Intell. (2012)
18. Osborne, M.J., Rubinstein, A.: Bargaining and Markets (Economic Theory, Econometrics, and Mathematical Economics). Academic Press (1990)
19. Rubinstein, A.: Perfect equilibrium in a bargaining model. Econometrica **50**(1), 97–109 (1982)
20. Rubinstein, A.: A bargaining model with incomplete information about time preferences. Econometrica **53**(5), 1151–1172 (1985)
21. Williams, C.R., Robu, V., Gerding, E.H., Jennings, N.R.: Using Gaussian processes to optimise concession in complex negotiations against unknown opponents. In: Proceedings of the 22nd International Joint Conference on Artificial Intelligence (IJCAI-2011), pp. 432–438 (2011)
22. Wurman, P.R., Wellman, M.P., Walsh, W.E.: The Michigan Internet AuctionBot: a configurable auction server for human and software agents. In: Proceedings of the Second International Conference on Autonomous Agents, pp. 301–308. ACM (1998)

Chapter 14
Collective Learning and Information Diffusion for Efficient Emergence of Social Norms

Chao Yu, Zhen Wang, Hongtao Lv, Honglin Bao and Yapeng Li

Abstract Social norms are believed to be the main cause of evolution and establishment of many complex systems in human societies, ranging from language lexicon systems to cultural codes of conduct. Revelation of mechanisms behind the emergence of social norms can not only provide us with a better understanding of formation and evolution processes of opinions, conventions and rules in human societies, but more importantly enable us to build and control large-scale complex systems. In this paper, a theoretical framework is proposed to study the emergence of social norms based on agent collective learning and information diffusion in complex relationship networks. In this framework, agents learn collectively from local interactions with their neighbors using multiagent learning methods, and diffuse their learnt information based on their underlying relationships. Extensive experiments are carried out to test the proposed framework in different topological and environmental settings and experimental results show that the framework is effective for emergence of social norms in complex relationship networks. The proposed framework emulates the opinion aggregation and knowledge transfer process in human and the research findings reveal some significant insights into efficient mechanisms of norm emergence in complex relationship networks.

C. Yu (✉) · H. Bao
School of Computer Science and Technology, Dalian University of Technology,
Dalian 116024, China
e-mail: cy496@dlut.edu.cn

H. Bao
e-mail: BHL19931025@mail.dlut.edu.cn

Z. Wang
School of Software, Dalian University of Technology, Dalian 116621, China
e-mail: wangzhen@ntu.edu.sg

H. Lv
School of Computer Engineering, Nanyang Technological University,
Singapore 639798, Singapore
e-mail: lvhongtao@mail.dlut.edu.cn

Y. Li
School of Innovation and Entrepreneurship, Dalian University of Technology,
Dalian 116024, China
e-mail: liyapeng@mail.dlut.edu.cn

Q. Bai et al. (eds.), *Multi-agent and Complex Systems*, Studies in Computational Intelligence 670, DOI 10.1007/978-981-10-2564-8_14

Keywords Norm emergence · Information diffusion · Learning · Multiagent systems

14.1 Introduction

One of the most critical problems in the coordinated control of large-scale MASs is to design efficient strategies that enable all the agents to reach an agreement in the areas of common interest. Social norms, which are used to control and regulate agent behaviors, have been considered to be an effective tool to facilitate coordination and cooperation among distributed agents. Therefore, recent years have seen a growing interest in social norms in MAS research and numerous theoretical investigations have been done under different assumptions, interaction protocols and network topologies [1–5].

It has been well recognized that learning from individual experience is a robust mechanism to facilitate emergence of social norms for distributed agents [6]. A great deal of work, therefore, has studied norm emergence either through agent learning from random interactions in a population of agents [1, 7, 8], or through agent learning from local interactions with neighbors by taking into consideration complex networks (e.g., small-world and scale-free networks) to model the underlying topology of the agent society [2, 9, 10]. Although these studies provide us with insights into some simple yet general mechanisms behind the emergence of social norms without a central controller, interaction protocols and learning settings in these studies are over simplified, thus are not capable to represent complex real-life situations. **First**, all these studies focus on studying norm emergence based on a simple interaction protocol, that is, each agent must be paired for interaction with one of its neighbors, randomly or preferentially, so that this agent can directly learn from the interaction. This interaction protocol simplifies real-life situations when individuals can collectively make a decision from multiple alternatives. This collective decision making is inherent in human nature because people often seek several opinions before making a final decision [11]. **Second**, all these studies focus on studying norm emergence based on agent learning from its own experience. This kind of individual learning indicates that each agent must interact with another agent so that the agent can directly learn from this interaction. In real life, however, people not only can learn from their individual trial-and-error experiences, but also can learn from the information directly provided by others. How to integrate social learning strategies into existing learning frameworks in order to facilitate norm emergence has not been well addressed in the current research. **Last**, all the previous studies considering the underlying topology among agents are situated in a singular relationship network. In the real world, however, interactions among people are usually governed by some complex relationships [12]. Actually, relationships in social networks reflect the direction of opinion influence and information diffusion among individuals, and thus are fundamental in determining the process of norm emergence.

Against this background, this paper proposes a general learning framework to study the emergence of social norms in complex relationship networks. In this framework, norms evolve as agents learn collectively over repeated interactions with their neighbors using multiagent reinforcement learning algorithms [13]. At each time step, an agent chooses a best-response action for each of its neighbors and aggregates all of these actions into an overall action using a number of opinion aggregation methods. The agent then plays the aggregated action with its neighbors and receives a corresponding reward towards each neighbor. Then, the learning information regarding each neighbor will be updated using the corresponding reward. To realize social learning in the framework, before next round play, agents communicate with their neighbors in order to exchange and update their learning information based on their underlying relationships. In this way, learnt information can be efficiently diffused among the agents. We investigate whether this kind of information diffusion can facilitate the process of norm emergence, and how a number of key issues such as neighborhood size and network topologies can influence norm emergence under the proposed learning framework. Research findings reveal some significant insights into the emergence process of social norms achieved by local collective behaviors and information diffusion among agents.

The remainder of the paper is organized as follows. Section 14.2 describes basic definitions. Section 14.3 presents the proposed learning framework. Section 14.4 gives the experimental studies. Section 14.5 discusses related work. Finally, Sect. 14.6 concludes the paper with some directions for future research.

14.2 Complex Relationship Networks and Social Norms

This section gives descriptions of complex relationship networks and social norms.

Definition 1 A **complex relationship network** can be represented as a directed graph $G = (E, R)$, where $E = \{e_1, \ldots, e_n\}$ is a set of entities (agents), and $R \subseteq E \times E$ represents a set of relationships, each of which connects two agents.

Two basic relationships are defined in the complex relationship networks, which are peer-to-peer relations and subordinate-superior relations, respectively.

Definition 2 A **peer-to-peer relation** $r_{i \approx j} = \langle e_i, e_j \rangle$ with $e_i \in E$, $e_j \in E$ and $r_{i \approx j} \in R$ is to indicate that agent e_i and agent e_j are peers and neither has a superior position or higher influence over the other.

Definition 3 A **subordinate-superior relation** $r_{i \preccurlyeq j} = \langle e_i, e_j \rangle$, $e_i \in E$, $e_j \in E$ and $r_{i \preccurlyeq j} \in R$ is to indicate that agent e_i is a subordinate of agent e_j, and thus is more inclined to be influenced by agent e_j.

Definition 4 Given a complex relationship network (E, R), the **neighbors** of agent i, denoted as $N(i)$, are a set of agents so that $N(i) = \{e_j \mid \langle e_i, e_j \rangle \in R\}$.

This paper focuses primarily on two typical topologies to represent a complex relationship network. They are small-world networks and scale-free networks. (1) **Small-world networks**. This kind of network is to represent the small-world phenomenon in many natural, social, and computer networks, where each node has only a small number of neighbors, and yet can reach any other node in a small number of hops. Small-world networks feature a high clustering coefficient and a short average path length. This kind of networks appears in many social networks such as the collaboration networks of film actors and academic researchers, and the friendship networks of high school students [14]. We use $SW_N^{k,\rho}$ to denote a small-world network, where k is the average size of the neighborhood of a node, ρ is the re-wiring probability to indicate the different orders of randomness of the network, and N is the number of nodes. (2) **Scale-free networks**. This kind of network is characterized by the power law of degree distribution of nodes, which means that a few "rich" nodes have high connectivity degrees, while the remaining nodes have low connectivity degrees. The probability that a node has k neighbors is roughly proportional to $k^{-\gamma}$. Examples of scale-free networks include the network of citations of scientific papers, and links between web pages on the World Wide Web [14]. These networks exhibit the feature of "preferential attachment", which means that the likelihood of connecting to a node depends on the connectivity degree of this node. We use $SF_N^{k,\gamma}$ to denote a scale-free network, in which N is the number of nodes.

This paper adopts learning "rules of the road" [15] as a metaphor to study the emergence of norms. In this scenario, agents strive to establish a convention/law of driving either on the left (L) side or on the right (R) side of the road. This interaction can be viewed as a pure coordination game displayed in Table 14.1.

The coordination game has two pure Nash-equilibria: both agents choose left or both agents choose right. Classical game theory, however, does not give a coherent account of how people would play a game like this. The problem is that there is nothing in the structure of the game itself that allows the players (even purely rational players) to infer what they ought to do. In reality, people can play such games because they can rely on some contextual cues to agree on a particular equilibrium [15]. One such contextual cue is social norms that can be used to guide and control people's behaviors towards specific ones.

Table 14.1 Payoff matrix of the pure Coordination Game

	Left (L)	Right (R)
Left (L)	1, 1	−1, −1
Right (R)	−1, −1	1, 1

14.3 The Proposed Framework

The sketch of the proposed learning framework is given by Algorithm 1. All agents in the society interact repeatedly and simultaneously with all their neighbors. At each time step, an agent chooses a best-response action for each neighbor using specific learning strategies (which will be described in detail in Sect. 14.3.1). The actions for all the neighbors are then aggregated to an overall action using a number of opinion aggregation methods (which will be described in detail in Sect. 14.3.2). The agent then plays the aggregated action with all of its neighbors and receives a corresponding reward. The learning information for each neighbor will then be updated by using the aggregated action and the corresponding reward. To realize information diffusion in the framework, before next round play, agents exchange and update their learning information based on their underlying relationships by means of communication (which will be explained in Sect. 14.3.3). The proposed collective learning framework is significantly different from all the frameworks for norm emergence in previous studies [2, 3, 10] by considering collective learning behaviors and information diffusion process in complex relationship networks. The final decision of an agent is affected by two kinds of learning processes: individual learning based on collective interaction experience with its neighbors and social learning achieved by exchanging and updating its learning experience based on the underlying relationships. These two learning processes interplay with each other, and therefore, can have a significant influence on norm emergence in the whole society.

Algorithm 1: The proposed learning framework

Initialize network and learning parameters;
for *each step t (t = 1, ..., T)* **do**
    for *each agent i (i = 1, ..., n)* **do**
        for *each neighbor* $j \in N(i)$ *of agent i* **do**
            Agent i chooses a best-response action $a_{i \to j}$ regarding agent j using a learning policy with exploration;
        end
        Agent i aggregates all the actions $a_{i \to j}$ into action a_i using ensemble learning methods;
    end
    for *each agent i (i=1, ..., n)* **do**
        Agent i plays action a_i with its neighbors and receives reward r_i^j for each interaction;
        Agent i updates learning information towards each neighbor using action-reward pair (a_i, r_i^j);
        Agent i communicates with each of its neighbors and updates its learning information based on their relationships.
    end
end

14.3.1 Learning Strategies for Interactions

Different learning strategies can be used for the interaction with each neighbor to determine the best-response action. This research focuses on reinforcement learning (RL) approaches [13], in which an agent learns a policy through trial-and-error interactions with its environment so as to maximize the expected discounted reward for each state in a sequential decision-making problem. One of the most important and widely used RL approach is Q-learning [16], in which an agent makes a decision through the estimation of a set of Q values. Its one-step updating rule is given by Eq. (14.1).

$$Q(s, a) \leftarrow Q(s, a) + \alpha[R(s, a) + \lambda \max_{a'} Q(s', a') - Q(s, a)] \tag{14.1}$$

where $\alpha \in (0, 1]$ is a learning rate, $\lambda \in [0, 1)$ is a discount factor, $R(s, a)$ and $Q(s, a)$ are the immediate and expected reward of choosing action a in state s at time step t, respectively, and $Q(s', a')$ is the expected discounted reward of choosing action a' in state s' at time step $t + 1$. $\max_{a'} Q(s', a')$ indicates the maximum value of the expected discounted reward in the new state s' at time step $t + 1$.

Every Q value of a state-action pair can be stored in a table for a discrete state-action space. In a state s, an agent chooses the best-response action a based on the Q values. The agent then transits to a new state s' and receives a reward $R(s, a)$ from the environment. The agent then updates the Q value of state-action pair (i.e., $Q(s, a)$) according to Eq. (14.1). It is proved that this tabular Q-learning method converges to the optimal $Q^*(s, a)$ w.p.1 when all state-action pairs are visited infinitely and an appropriate learning rate is chosen [16].

During learning, an agent needs to make a balance between the exploitation of learnt knowledge and the exploration of unexplore environment in order to learn an optimal policy [13]. The ε-greedy exploration policy is an efficient mechanism to trade off exploitation and exploration during learning, which can be given by Eq. (14.2) [13].

$$\pi(s, a) = \begin{cases} 1 - \varepsilon & \text{if } a = \arg\max_{a'} Q(s, a'), \\ \varepsilon & \text{otherwise.} \end{cases} \tag{14.2}$$

where $\varepsilon \in [0, 1]$ is the exploration rate.

Equation (14.2) means that an agent chooses the action with the highest Q-value with a probability of $1 - \varepsilon$ (i.e., exploitation of the learnt knowledge) and chooses other actions randomly with a probability of ε (i.e., exploration of the unknown environment). The value of ε is often set small to indicate a small probability of exploration.

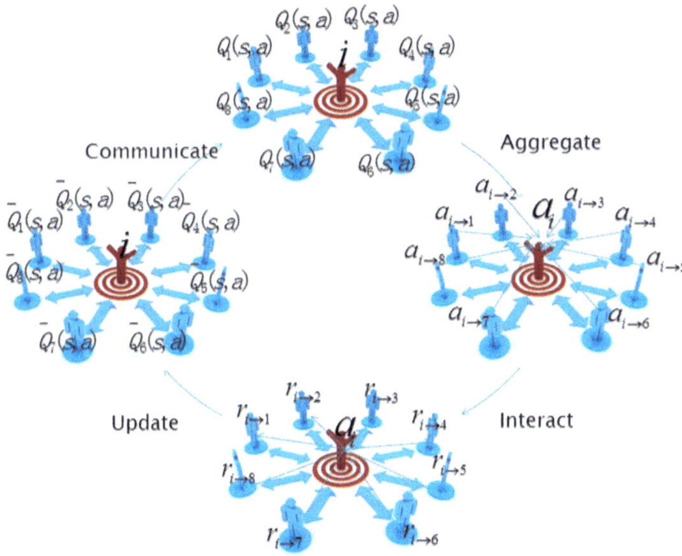

Fig. 14.1 Illustration of the proposed framework

Figure 14.1 gives a vivid illustration of the proposed learning framework based on Q-learning. Each focal agent i keeps learning information in terms of a Q-value table $Q_j(s, a)$ for each neighbor j. At each time step, regarding each neighbor j, agent i chooses the best-response action with the highest Q-value based on the corresponding Q-value table with a probability of $1 - \varepsilon$ (i.e., exploitation), or chooses other actions randomly with a probability of ε (i.e., exploration). Agent i then collectively makes a decision by aggregating all the actions $a_{i \to j}$ for each neighbor into a final action a_i and interacts with each of its neighbors using action a_i. The Q-value table $Q_j(s, a)$ regarding neighbor j then can be updated to $\overline{Q_j}(s, a)$ according to Equation (14.1) after the agent receives the immediate reward $r_{i \to j}$. Finally, agent i communicates with each of its neighbors to update $\overline{Q_j}(s, a)$ to $Q_j(s, a)$ used for next round play.

14.3.2 Opinion Aggregation Methods

In the proposed framework, each agent needs to aggregate the best-response actions regarding its neighbors into a final action. This models human's opinion aggregation process in that people usually consult with many others before making a final decision. The opinion aggregation process can be realized by using some ensemble learning methods, which are traditionally used to weigh several individual classifiers and then combine them in order to make a final decision that will be better than the one made by each of them separately [11]. In [3], we proposed a general collective learning framework for norm emergence, in which the agent combines all the best-response actions regarding each neighbor to make a final decision by considering each neighbor's position (e.g., degree of connectivity) as well as the neighbor's performance in past interactions.

A focal agent has its preference for each action choice. The selection of the final action is to choose the one with the highest preference. Two basic methods have been proposed to calculate the preference of each action. They are **Majority voting** and **Weighted voting**. (1) In the majority voting, the preference values are calculated by the majority voting ensemble method. The most preferred action is simply the one that is suggested by most of the neighbors. The principle of this method reflects the fact that people are social beings and can be influenced by each other so that people are more prone to accept the opinion that is adopted by the majority of their neighbors. (2) In the weighted voting, the best-response action regarding a neighbor is weighed by a weight considering that each neighbor in the network can occupy different positions, and thus can play a different role in shaping the norms of the whole society. Assume that the decision regarding neighbor j is weighed by weight $w_{t,j}$. Two different ways can be used to determine weights. The **Structure-based approach** considers the different structural positions of agents in the network. A straightforward way of defining the structure-based weight of each agent is to use the agent's degree of connectivity. The **Performance-based approach** determines each neighbor's weight according to past interaction experience between this neighbor and the focal agent. If a neighbor's action is always consistent with the agent's own action, the agent will then consider the neighbor to be more trustworthy and accordingly assign a higher weight to this neighbor. Weight $w_{t,j}$ can be given by Eq. 14.3:

$$w_{t,j} = w_{t-1,j} + \beta(s - w_{t-1,j}) \tag{14.3}$$

where $w_{0,j} = \frac{1}{|N(j)|}$; β is a learning rate; and $s = 1$ if interaction at time $t - 1$ is successful, otherwise $s = 0$. The interaction is successful if the interaction brings a positive reward to the agent, namely, the actions of the interacting agents are consistent with each other.

14.3.3 Information Diffusion Methods

Many strategies can be used to realize agent social learning process. This paper adopts the simplest form of social learning by assuming communication among agents. After receiving the reward from the environment and updating the learning information, e.g., Q-values, each agent then communicates with its neighbors and exchanges their learning information based on their underlying relationships.

Algorithm 2: The information diffusion methods

// $Q_{i \to j}$: agent $i's$ Q-values regarding to its neighbor j;
for *each neighbor* $j \in N(i)$ *of agent i* **do**
    if $r_{i \approx j}$ **then**
        $Q_{i \to j} \Leftarrow (Q_{i \to j} + Q_{j \to i})/2$;
        $Q_{j \to i} \Leftarrow (Q_{i \to j} + Q_{j \to i})/2$;
    end
    if $r_{i \preccurlyeq j}$ **then**
        $Q_{i \to j} \Leftarrow Q_{j \to i}$;
    end
    if $r_{i \succcurlyeq j}$ **then**
        $Q_{j \to i} \Leftarrow Q_{i \to j}$;
    end
end

Algorithm 2 shows how a focal agent i communicates with its neighbors. The idea is as follows: if agent i is a peer of neighbor j, since neither side is able to gain dominant influence on the other, agent i and neighbor j then synthesize their learning experience by averaging their Q-values; if agent i is a subordinate of neighbor j, agent i then takes neighbor j' Q-values as its own value because it is more inclined to influenced by neighbor j; on the contrary, if agent i is superior to neighbor j, agent i communicates its Q-values to neighbor j for value updating but keeps its own Q-values unchanged. Although this kind of communication is very simple, it provides an efficient means to reflect opinion influence and information diffusion among agents based on their underlying relationships.

14.3.4 Global Exploration Mode

In Algorithm 1, the exploration process is conducted during an agent's local interactions with each of its neighbors. This means that the agent conducts exploration for each neighbor before aggregating all the actions into a final deterministic action. We call this kind of exploration *local exploration* mode. In order to study the impact of different levels of uncertainties caused by exploration on the learning performance, we also propose another exploration mode in Algorithm 3, in which an agent determines a greedy action using a learning strategy regarding each of its neighbors (Line 5), and then aggregates these actions for all the neighbors into an overall action a_i' (Line 7). Exploration is then conducted when the agent chooses the final action a_i based on the aggregated overall action a_i' (Line 8). We call this kind of exploration *global exploration* mode.

Algorithm 3: The learning framework with global exploration mode

Initialize network and learning parameters;
for *each step t (t = 1, ..., T)* **do**
  for *each agent i (i = 1, ..., n)* **do**
    for *each neighbor* $j \in N(i)$ *of agent i* **do**
      Agent i chooses greedy action $a_{i \to j}$ regarding neighbor j using a learning strategy;
    end
    Agent i aggregates all the actions $a_{i \to j}$ into action a_i';
    Agent i chooses final action a_i based on a_i' with exploration;
  end
  for *each agent i (i = 1, ..., n)* **do**
    for *each neighbor* $j \in N(i)$ *of agent i* **do**
      Agent i plays action a_i with neighbor j and receives corresponding reward $r_{i \to j}$;
      Agent i updates learning information regarding neighbor j using $\langle a_i, r_{i \to j} \rangle$;
      Agent i communicates with each of its neighbors and updates its learning information based on their relationships.
    end
  end
end

14.4 Experimental Studies

The purpose of this experiment is to study the emergence of social norms in the proposed learning framework. The performance metrics are the convergence ratio of the social norms (i.e., how many agents in the society can reach a final consensus) and the time needed to reach such a consensus. A social norm is said to be established when at least 90 % of the agents have adopted the same action. We let $T_{\%90}$ denote the convergence time when such a norm emerges.

14.4.1 Experimental Settings

The Watts-Strogatz model [17] is used to generate a small-world network, and the Barabasi-Albert model [14] is used to generate a scale-free network. To initialize the relationship networks, the percentages of superiors, subordinates and peers are set to equal (i.e., accounting for a third of the population, respectively). Learning rate α in Q-learning is 0.1, and ε in the ε-exploration strategy is 0.1 in order to indicate a small probability of exploration. Learning rate β in the performance-based weighted

Table 14.2 Parameter settings for the experiments

Parameters in the experiments			
Categories	Parameters	Values	Meanings
Network	N	[50, 1000]	number of agents
	k	{2, 4, 6, 12, 14, 20}	average size of neighborhoods in small-world networks
	ρ	{0, 0.2, 0.4, 0.6, 0.8, 1}	re-wiring probability in small-world networks
	m_0	2	initial agents to generate scale-free networks
	l	1	edge added to the network at every time step in small-world networks
	γ	3	the exponent value of power law in scale-free networks
Learning algorithm	α	0.1	learning rate in Q-learning
	ε	0.10	ε-exploration rate
Others	β	0.8	learning rate to adjust the weight
	N_a	{4, 6, 10, 20}	number of actions available to the agents

method is 0.8. Number of optional behavior is defined as 4. In this study, unless stated otherwise, we use the small-world network as the default network topology due to the variety of this kind of network, local exploration as the exploration mode, and the majority voting as the opinion aggregation method. All results are averaged over 100 independent Monte-Carlo runs. The parameter settings in the experimental study are summarized in Table 14.2 for clarity.

14.4.2 Results and Analysis

Influence of Communication Previous work [3] has shown that a social norm can always emerge when agents learn collectively through local interactions with neighbors and this kind of collective learning can be more efficient and robust than the paired-learning framework. It is not clear, however, whether a social norm can still successfully emerge in the whole society when considering social learning during agents' collective decision making in complex relationship networks. If a social norm does emerge, what is the ratio of such an emergence? What is the influence of information diffusion (which is realized assuming communication among agents based on their relationships) on the efficiency of norm emergence?

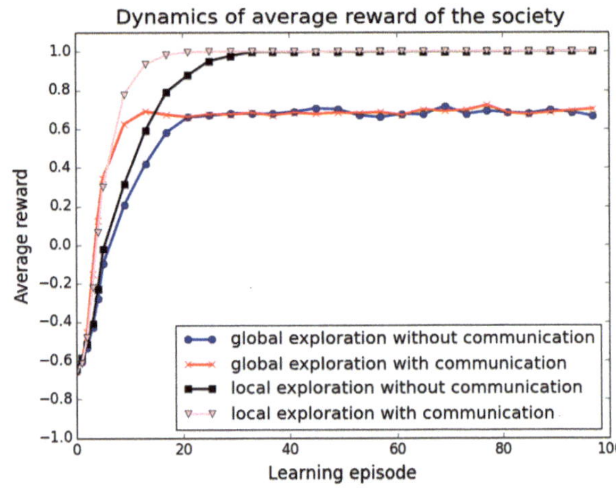

Fig. 14.2 Influence of information diffusion (in terms of communication)

To answer these questions, we first test the proposed framework in small-world network $SW_{100}^{12,0.8}$, and the learning dynamics of the average reward of the whole population is plotted in Fig. 14.2, where the strategy without communication means that agents learn collectively from local interactions but do not diffuse their learning information based on their relationships. From the results, we can see that both strategies with and without communication can emerge social norms successfully. As the learning process moves on, the number of agents who choose the same action as the norm increases. This means that more and more agents have reached a consensus on which action should be the norm, and this consensus correspondingly increases the average payoff dramatically. A norm, however, emerges faster under the collective learning framework with communication than under the collective learning framework without communication. This is because agents using collective learning framework with communication can promote the information exchange among agents and therefore decrease the diversity of the society. These results indicate the merits of information diffusion in our framework and further confirm that information exchange through social learning can facilitate emergence of social norms. Figure 14.2 also shows that the local exploration mode outperforms the global exploration mode. This is because agents using local exploration mode explore the environment locally and make a final decision collectively, so that the uncertainties caused by the exploration decrease.

Figure 14.3 shows the frequency of each action adopted by the agents when norm (action 0) emerges in the population. Initially, each agent randomly chooses an action, so there are respectively about 25 % of the agents to choose each action. As the learning process moves on, however, the number of agents who choose action 0 as the norm increases rapidly, and finally reaches about 92 % in the population with global exploration and 100 % with local exploration. This means that more and more agents have reached a consensus on that action 0 should be the norm, and this consensus correspondingly increases the average payoff dramatically.

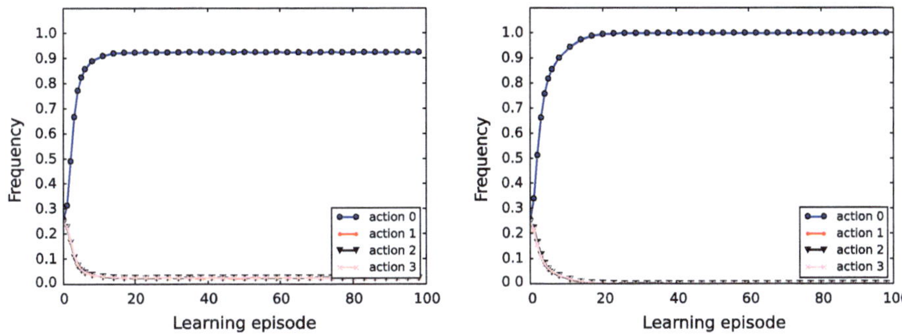

(a) dynamics with global exploration (b) dynamics with local exploration

Fig. 14.3 Action dynamics with different exploration modes in norm emergence

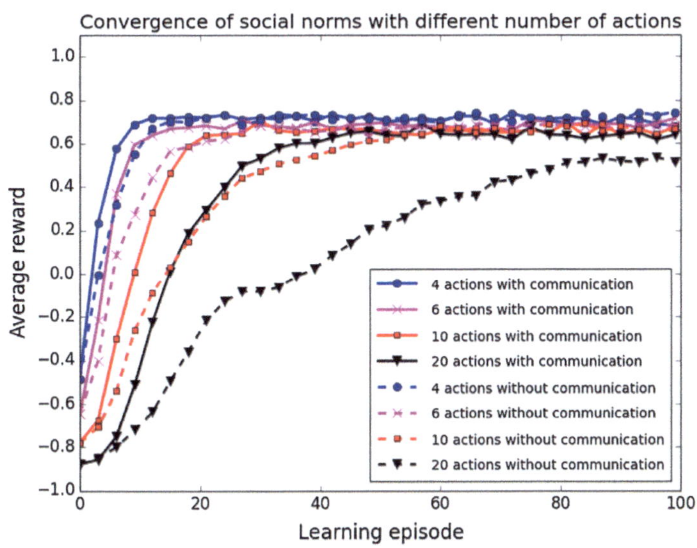

Fig. 14.4 Influence of number of actions

Influence of Number of Actions It has been shown that the number of actions available to the agents is an important factor in the emergence of social norms [1, 3, 10]. Thus, we vary the number of actions in the set $N_a = \{4, 6, 10, 20\}$ in network $SW_{100}^{12,0.8}$ to investigate its impact on norm emergence. Only when two agents choose the same action they will receive a payoff of 1. Otherwise, they receive a payoff of -1. Results in Fig. 14.4 show that a larger number of available actions causes a delayed convergence of norms. This is because a larger number of actions may produce more varied local sub-norms, leading to diversity across the society. It thus takes a longer time for the agents to eliminate this diversity to achieve a final consensus, and thus norm emergence is prolonged throughout the network. Figure 14.4 also shows

Table 14.3 Speed of norm emergence with different methods

Speed ($T_{90\,\%}$)	4 actions	6 actions	10 actions	20 actions
Local exploration with communication	4.0	6.0	10.5	24.9
Global exploration with communication	5.3	8.4	13.6	26.6
Local exploration without communication	6.4	10.7	21.5	63.1
Global exploration without communication	10.9	19.6	42.4	90.7

that in all cases of action numbers, social norms emerge slower when agents do not communicate their learning experience with each other. These results further confirm that information diffusion among agents can facilitate the emergence of social norms in complex relationship networks.

Table 14.3 shows the number of time steps needed for 90 % of the agents to choose the same action as a social norm. We can see that with the increase of the number of actions, the acceleration of emergence using communication is getting more and more prominent. This means that our method considering information diffusion among agents is more suitable for large action spaces. Moreover, it is revealed that the collective learning with local exploration performs better than that with global exploration, especially when the action spaces are large.

Influence of Neighborhood Sizes and Network Randomness The number of neighbors and network randomness are another two important factors that can influence the emergence of social norms [1, 9]. Figure 14.5 shows the dynamics of average agent reward with different neighborhood size k in network $SW_{100}^{k,0.8}$, from which we can see that when the average number of neighbors increases, the speed of norm emergence is reduced. This effect is due to the clustering coefficient of the network. Clustering coefficient is used to measure the degree to the nodes in a graph which tend to cluster together. When the number of neighbors increases, the clustering coefficient increases accordingly, therefore agents located in different positions only need a smaller number of interactions to reach a consensus. On the other hand, when agents have a small neighborhood size, they only interact with their neighbors, which account for a small part of the whole population. This results in diverse subnormal norms formed at different regions of the network.

Figure 14.6 shows the influence of network randomness on norm emergence in network $SW_N^{k,\rho}$. When $\rho = 0$, network $SW_N^{k,\rho}$ is reduced to a regular ring lattice. Increasing rewiring probability ρ produces a small-world network with increasing randomness. When $\rho = 1$, the network becomes a fully random network. The results indicate that it is more efficient for a norm to emerge in a network with higher randomness. This is because the increase in randomness can reduce the network diameter (i.e., the largest number of hops in order to traverse from one vertex to another [9]), and the smaller a network diameter is, the more efficient for the network to evolve a social norm [18].

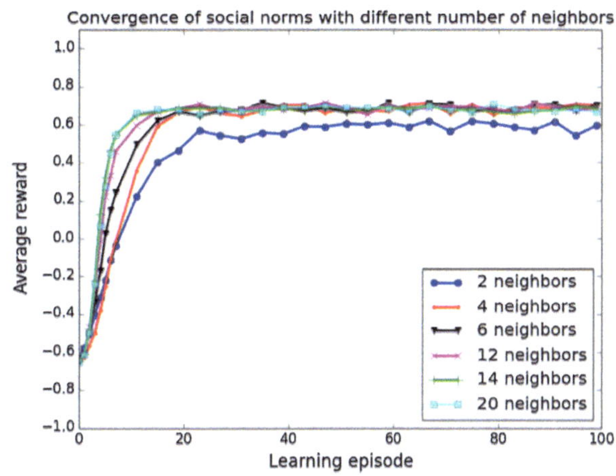

Fig. 14.5 Influence of neighborhood sizes

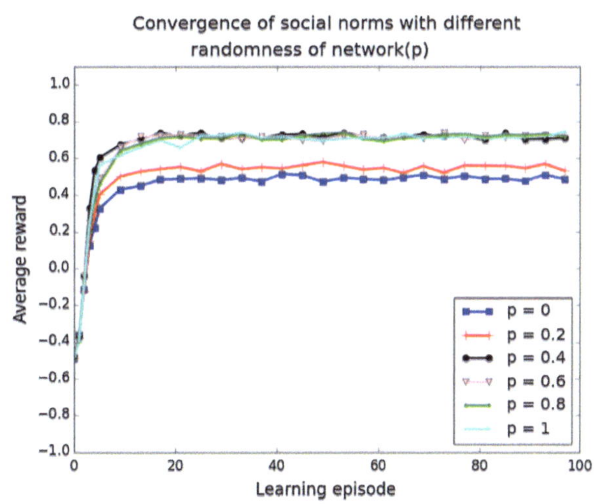

Fig. 14.6 Influence of network randomness

14.4.2.1 Influence of Network Topologies and Opinion Aggregation Methods

The learning dynamics with different network topologies is shown in Fig. 14.7, from which we can see that various network topologies can have significant impacts on the emergence of social norms. The gird network and the scale-free network can bring about more efficient emergence of social norms than the small-world network. This result is in accordance with the reported results in various previous studies [18, 19]. The influence of different opinion aggregation methods on norm emergence in small-world network ($SW_{100}^{12,0.1}$) is shown in Fig. 14.8. As can be seen, the performance-based method performs best among the three methods, followed by the majority

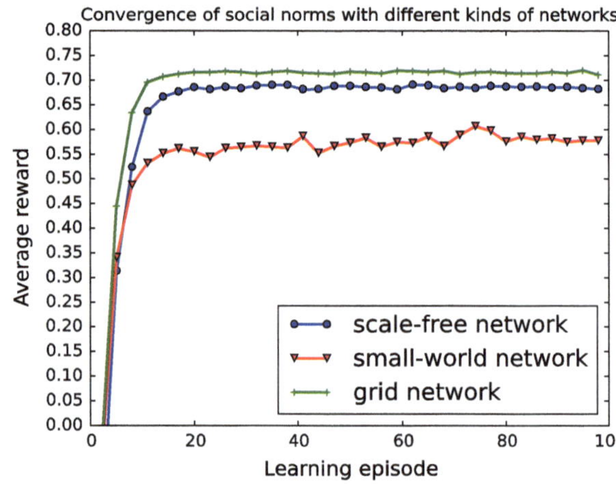

Fig. 14.7 Influence of neighborhood sizes

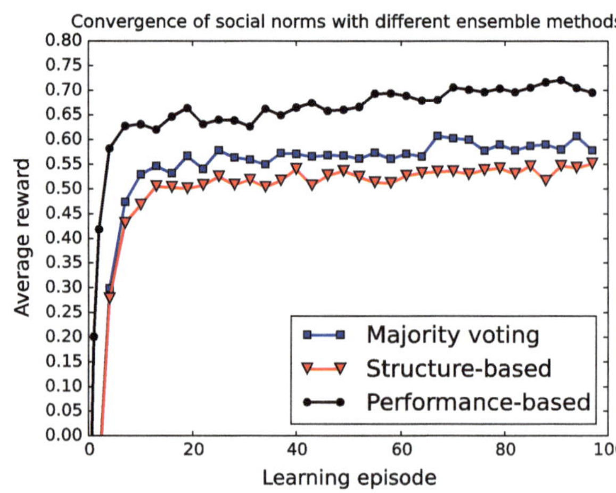

Fig. 14.8 Influence of opinion aggregation methods

voting method and then the structure-based method. This result demonstrates the merits of considering the trust of neighbors during the collective learning process in facilitating emergence of social norms.

14.5 Related Work

The emergence of social norms has gained increasing attention in the area of MASs. Shoham and Tennenholtz [20] proposed an approach based on the Highest Cumulative Reward (HCR) rule to study the emergence of social norms. Sen et al. [1, 7] proposed a social learning mechanism for norm emergence. Other works have studied norm emergence by examining the underlying network topology of agents. Sen et al.

[10] evaluated how varying topologies of social networks affected the emergence of norms through social learning. Villatoro et al. [9] investigated the effects of memory and the history of past activities during learning on the success and rate of emergence of social norms in different network structures. All these previous studies did not consider social learning and complex relationships among agents during norm emergence. The focus of this paper, however, is to investigate whether the emergence of social norms can be facilitated by exchanging learning experience based on the underlying relationships among agents. This focus differentiates our work from all these previous studies.

Several studies also investigated the impact of social learning on norm emergence. For example, Savarimuthu et al. [8] demonstrated the usefulness of combining both individual learning (i.e., experiential) and two other forms of social learning (i.e., observational and communication-based learning) to boost the convergence of social norms; and Villatoro et al. [2] used social learning (i.e., observation) as an efficient social instrument to effectively address the *frontier effect* problem so as to facilitate norm emergence in networked agent societies. More recently, Yu et al. [21] studied the role of social learning in norm emergence, and showed that social learning was less valuable than individual learning in dynamic environments. In all these studies, however, complex relationships among agents are not considered and no information diffusion is caused by social learning. This is against our work, in which social learning is adopted for agents to diffuse their learning information based on their underlying relationships. The focus of our work is to investigate whether such kind of information diffusion can be helpful to facilitate emergence of social norms.

14.6 Conclusion and Future Work

Norm emergence has been extensively studied in MASs. The existing work in this area, however, has mainly neglected three important factors in norm emergence, i.e., agents' collective behaviors, agents' capability of social learning and the complex relationships among agents. This paper conquers the above shortcomings by proposing a learning framework for norm emergence in complex relationship networks, based on collective learning and information diffusion among agents. The goal of this work is to investigate whether norm emergence can be facilitated by information diffusion using social learning based on the underlying relationships among agents. Extensive experiments are carried out to test the proposed framework in different topological and parameter settings and experimental results show that it is indeed effective for a social norm to emerge in complex relationship networks under the proposed learning framework.

This paper is an initial step towards the goal of examining and developing robust mechanisms that can facilitate norm emergence in more complex and realistic social situations. To realize this goal, much work still remains to be done. For example, relationships between agents can be adjusted dynamically during interactions. In addition, relation strength can be added into the definition of relationships to indicate how strong a relation is between two agents.

Acknowledgments This work is supported by the National Natural Science Foundation of China under Grant 61502072, Fundamental Research Funds for the Central Universities of China under Grant DUT14RC(3)064), and Post-Doctoral Science Foundation of China under Grants 2014M561229 and 2015T80251.

References

1. Sen, S., Airiau, S.: Emergence of norms through social learning. In: Proceeding of 20nd IJCAI, pp. 1507–1512 (2007)
2. Villatoro, D., Sabater-Mir, J., Sen, S.: Social instruments for robust convention emergence. In: Proceedings of 22nd IJCAI, pp. 420–425 (2011)
3. Yu, C., Zhang, M., Ren, F., Luo, X.: Emergence of social norms through collective learning in networked agent societies. In: Proceedings of 12th AAMAS, pp. 475–482 (2013)
4. Yu, C., Zhang, M., Ren, F.: Collective learning for the emergence of social norms in networked multiagent systems. IEEE Trans. Cybern. **44**(12), 2342–2355 (2014)
5. Hao, J., Sun, J., Huang, D., Cai, Y., Yu, C.: Heuristic collective learning for efficient and robust emergence of social norms. In: Proceedings of 14th AAMAS, pp. 1647–1648 (2015)
6. Savarimuthu, B., Cranefield, S.: Norm creation, spreading and emergence: a survey of simulation models of norms in multi-agent systems. Multi-agent Grid Syst. **7**(1), 21–54 (2011)
7. Mukherjee, P., Sen, S., Airiau, S.: Norm emergence under constrained interactions in diverse societies. In: Proceedings of 7th AAMAS, pp: 779–786 (2008)
8. Savarimuthu, B., Arulanandam, R., Purvis, M.: Aspects of active norm learning and the effect of lying on norm emergence in agent societies. In: Agents in Principle, Agents in Practice. LNAI, vol. 7047, pp. 36–50 (2011)
9. Villatoro, D., Sen, S., Sabater-Mir, J.: Topology and memory effect on convention emergence. In: 2009 IEEE/WIC/ACM International Joint Conference on Web Intelligence and Agent Technology, pp. 233–240 (2009)
10. Airiau, S., Sen, S.: Effects of social network topology and options on norm emergence. Lecture Notes on Computer Science, pp. 211–222 (2010)
11. Polikar, R.: Ensemble based systems in decision making. IEEE Circ. Syst. Mag. **6**(3), 21–45 (2006)
12. Hanneman, R., Riddle, M.: Introduction to Social Network Methods. University of California Riverside (2005)
13. Sutton, R.S., Barto, A.G.: Reinforcement Learning: An Introduction. MIT press, Cambridge (1998)
14. Barabási, A.L., Albert, R.: Statistical mechanics of complex networks. Rev. Modern Phys. **74**, 47–97 (1998)
15. Young, H.P.: The economics of convention. J. Econ. Perspect. 105–122 (1996)
16. Watkins, C., Dayan, P.: Q-learning. Mach. Learn. **8**(3–4), 279–292 (1992)
17. Watts, D., Strogatz, S.: Collective dynamics of small-world networks. Nature **393**(6684), 440–442 (1998)
18. Delgado, J.: Emergence of social conventions in complex networks. Artif. Intell. **141**(1), 171–185 (2002)
19. Shibusawa, R., Sugawara, T.: Norm emergence via influential weight propagation in complex networks. In: 2014 European Network Intelligence Conference, pp. 30–37 (2014)
20. Shoham, Y., Tennenholtz, M.: On the emergence of social conventions: modeling, analysis, and simulations. Artif. Intell. **94**(1), 139–166 (1997)
21. Yu, C., Zhang, M., Ren, F., Hao, J.: Emergence of social norms through collective learning in networked agent societies. In: Proceedings of MFSC@AAMAS2014 (2014)

MIX
Papier aus verantwortungsvollen Quellen
Paper from responsible sources
FSC® C105338

If you have any concerns about our products,
you can contact us on
ProductSafety@springernature.com

In case Publisher is established outside the EU,
the EU authorized representative is:
Springer Nature Customer Service Center GmbH
Europaplatz 3, 69115 Heidelberg, Germany

Printed by Libri Plureos GmbH
in Hamburg, Germany